NEBOSH ENVIRONMENTAL MANAGEMENT CERTIFICATE

UNIT EMC1: ENVIRONMENTAL MANAGEMENT AND UNIT EMC2: ASSESSING ENVIRONMENTAL ASPECTS AND ASSOCIATED IMPACTS

Element 1: Foundations in Environmental Management

Element 2: Environmental Management Systems

Element 3: Assessing Environmental Aspects and Impacts

Element 4: Planning for and Dealing with Environmental Emergencies

Element 5: Control of Emissions to Air

Element 6: Control of Environmental Noise

Element 7: Control of Contamination of Water Sources

Element 8: Control of Waste and Land Use

Element 9: Sources and Use of Energy and Energy Efficiency

Unit EMC2: Assessing Environmental Aspects and Impacts Practical Assessment

Contributors

John Binns, BSc (Hons), MSc, MSc, MIEMA

© RRC International

All rights reserved. RRC International is the trading name of The Rapid Results College Limited, Tuition House, 27-37 St George's Road, London, SW19 4DS, UK.

These materials are provided under licence from The Rapid Results College Limited. No part of this publication may be reproduced, stored in a retrieval system, or transmitted in any form, or by any means, electronic, electrostatic, mechanical, photocopied or otherwise, without the express permission in writing from RRC Publishing.

For information on all RRC publications and training courses, visit:
www.rrc.co.uk

RRC Module No: EMC1

ISBN for this volume: 978-1-912652-46-4

Third edition August 2022

ACKNOWLEDGMENTS

RRC International would like to thank the National Examination Board in Occupational Safety and Health (NEBOSH) for their co-operation in allowing us to reproduce extracts from their syllabus guides.

This publication contains public sector information published by the Health and Safety Executive and licensed under the Open Government Licence v.3 (www.nationalarchives.gov.uk/doc/open-government-licence/version/3).

Every effort has been made to trace copyright material and obtain permission to reproduce it. If there are any errors or omissions, RRC would welcome notification so that corrections may be incorporated in future reprints or editions of this material.

Whilst the information in this book is believed to be true and accurate at the date of going to press, neither the author nor the publisher can accept any legal responsibility or liability for any errors or omissions that may be made.

Element 1: Foundations in Environmental Management

The Scope and Nature of Environmental Management	**1-3**
Definition of the Environment	1-3
The Multidisciplinary Nature of Environmental Management	1-4
Size of the Environmental Problem	1-5
The Ethical, Legal and Financial Reasons for Maintaining and Promoting Environmental Management	**1-19**
Rights and Expectations of Internal and External Interested Parties (Local Residents, Indigenous People, Supply Chain, Customers and Workers)	1-20
Outcomes of Incidents	1-21
The Actions and Implications of Pressure Groups	1-21
Overview of Compliance Issues	1-22
The Business Case for Environmental Management	1-24
Supporting Sustainable Development	**1-28**
Definition of Sustainability	1-28
Importance of Sustainable Development	1-28
The Business Case for Sustainable Development	1-31
Role of the United Nations' Sustainable Development Goals (SDGs)	1-32
The Role of National Governments and International Bodies in Formulating a Framework for the Regulation of Environmental Management	**1-34**
International Law	1-34
The Importance of Knowing and Understanding Local Legislation	1-39
Meaning of BAT and BPEO	1-39
The Role of Enforcement Agencies and Consequences of Non-Compliance	1-39
Summary	**1-41**
Exam Skills	**1-42**

Contents

Element 2: Environmental Management Systems

Reasons for Implementing an Environmental Management System (EMS) — 2-3
Introduction to Environmental Management Systems — 2-3

The Key Features and Appropriate Content of an Effective EMS — 2-6
Introduction to ISO 14001 — 2-6
ISO 14001 — 2-6
Initial Environmental Review — 2-7
Context of the Organisation — 2-8
Leadership — 2-9
Planning — 2-10
Support — 2-13
Operation — 2-14
Performance Evaluation — 2-15
Improvement — 2-24
Eco-Management and Audit Scheme (EMAS) — 2-25

Benefits and Limitations of Introducing a Formal EMS into the Workplace — 2-26
Benefits of Introducing a Formal EMS into an Organisation — 2-26
Limitations of Introducing a Formal EMS into an Organisation — 2-27

Summary — 2-29

Exam Skills — 2-30

Element 3: Assessing Environmental Aspects and Impacts

Reasons for Carrying Out Environmental Aspect and Impact Assessments	**3-3**
Why Identify Environmental Impacts?	3-3
Aims and Objectives of Impact Assessment	3-3
Life-Cycle Analysis (Cradle-to-Grave Concept)	3-6
Types of Environmental Impact	**3-8**
Direct and Indirect Impacts	3-8
Cumulative Impacts	3-8
Positive and Negative Effects	3-8
Nature and Key Sources of Environmental Information	**3-11**
Internal to the Organisation	3-11
External to the Organisation	3-12
Identifying Environmental Aspects and Associated Impacts	**3-13**
Implementing an EMS	3-13
Identifying Environmental Aspects	3-13
Determining Associated Environmental Impacts	3-14
Specific Impacts	3-16
Identifying Receptors at Risk	3-17
Identification of Aspects and Impacts	3-17
Evaluating Impact and Adequacy of Current Controls	3-19
Risk and Opportunities	3-20
Recording Significant Aspects and Impacts	3-21
Reviewing	3-21
Summary	**3-22**
Exam Skills	**3-23**

Contents

Element 4: Planning for and Dealing with Environmental Emergencies

The Importance of Environmental Emergency Planning	4-3
The Effects of Unplanned Incidents on the Environment	4-3
General Duty or Responsibility Not to Pollute	4-3
Requirement of the Environmental Management System	4-3
Need for Prompt Action to Protect People, the Environment and Organisational Assets	4-4
Immediate Risks	4-4
Long-Term Risks	4-4
Emergency Preparedness and Response	4-5
Recognising Risk Situations and Action to Take	4-5
Emergency Response Plans	4-7
Information and Training for Internal and External Interested Parties	4-9
Review and Continual Improvement of Emergency Response Plans	4-12
Summary	4-13
Exam Skills	4-14

Element 5: Control of Emissions to Air

Air Quality Standards	5-3
Why are there Air Quality Standards?	5-3
Meaning and Uses of ppm and mg/m^3	5-3
The Potential Effects of Poor Air Quality	5-3
The Role of Air Quality Standards	5-4
Main Types of Emissions to Atmosphere	5-5
Types of Emission and their Hazards	5-5
Sources of Air Pollution	5-6
Common Pollutants	5-8
Control Measures to Reduce Emissions	5-12
Controlling Air Pollution	5-12
Control Hierarchy	5-12
Examples of Technology	5-13
Unit EMC2: Environmental Practical Application	5-25
Summary	5-26
Exam Skills	5-27

Contents

Element 6: Control of Environmental Noise

Sources and Effects of Environmental Noise	**6-3**
The Characteristics of Noise which Lead to it Being a Nuisance	6-3
The Effects of Noise	6-4
Legal Considerations	6-5
Common Sources of Environmental Noise	6-5
Methods for the Control of Environmental Noise	**6-8**
Monitoring Requirements and Arrangements	6-8
Basic Noise Control Techniques	6-8
Management Controls	6-11
Summary	**6-12**
Exam Skills	**6-13**

Element 7: Control of Contamination of Water Sources

Importance of the Quality of Water for Life	**7-3**
The Meaning of Safe Drinking Water, Groundwater, Surface Water	7-3
The Water Cycle	7-5
Water for Agriculture and Industry	7-7
The Potential Effects of Water Pollution to the Environment	7-7
Over-Abstraction	7-8
Desalination	7-8
Water Conservation	7-8
The Potential Effects of Pollution on Water Quality	7-9
Main Issues and Impacts of Ocean Pollution	7-9
Main Sources of Water Pollution	**7-11**
Controlling Sources of Water Pollution	7-11
Main Control Measures Available to Reduce Contamination of Water Sources	**7-17**
Control Hierarchy	7-17
Monitoring Water Quality	7-18
Control Methods	7-19
Controls for Storage and Spillage	7-20
Controls for Wastewater	7-26
Difficulties in Maintaining Equipment in Some Locations or Environments	7-28
Summary	**7-29**
Exam Skills	**7-30**

Element 8: Control of Waste and Land Use

Waste Types — 8-3

The Waste Framework Directive — 8-3
Definition of Waste — 8-3
Inert Waste — 8-4
Hazardous Waste — 8-4
Non-Hazardous Waste — 8-5
Clinical Waste — 8-5
Radioactive Waste — 8-6
Waste Types Subject to Specific Legal Requirements — 8-6
Types of Waste — 8-6

Minimising Waste — 8-8

Impacts from Waste — 8-8
Waste - a Worldwide Problem — 8-8
The Business Case for Minimising Waste — 8-9
The Waste Hierarchy — 8-10
Applying the Waste Hierarchy at Every Stage — 8-11

Managing Waste — 8-13

What is the Waste Chain? — 8-13
Recognition of the Key Steps — 8-13
Responsible Waste Management — 8-13
Benefits, Limitations and Barriers to Re-Use and Recycling — 8-14
On-Site Separation and Storage including Segregation, Identification and Labelling — 8-15
Transportation including Transfer to an Authorised Person and Required Regulatory Documentation — 8-16
Differing Legal Requirements for Waste — 8-17
Disposal — 8-18
Producer Responsibility — 8-18
Packaging Waste — 8-18
Electrical and Electronic Waste — 8-18
Waste from Construction Projects — 8-20

Outlets Available for Waste — 8-21

Circular Economy — 8-21
Landfill and Incineration as Ultimate Disposal Routes — 8-22
Other Treatment or Disposal Routes — 8-25
Global Waste Trade — 8-26
Export Costs and the Impact of Export, Landfill and Aggregate Taxes — 8-26

Risks Associated with Contaminated Land — 8-28

The Potential Effects of Contaminated Land on the Environment — 8-28
Contaminated Land Liabilities — 8-29

Summary — 8-31

Exam Skills — 8-32

Contents

Element 9: Sources and Use of Energy Efficiency

Use of Fossil Fuels	9-3
Examples of Fossil Fuels	9-3
Benefits and Limitations of their Use as an Energy Source	9-6
Carbon Offsetting	9-7
Renewable Sources of Energy	9-9
Fossil Fuel Alternatives	9-9
Benefits and Limitations of the Use of Alternative Energy Sources	9-14
Energy Supply in Remote Locations and Developing Countries	9-16
On-Site Energy Generation and Storage	9-16
Benefits and Limitations of Using Emerging Technologies for Energy	9-16
Energy Efficiency	9-17
Benefits of Energy Efficiency	9-17
Energy Monitoring	9-17
Control Measures Available to Increase Energy Efficiency	9-18
Fuel Choice for Transport and the Optimisation of Vehicle Use	9-21
Summary	9-24
Exam Skills	9-25

Final Reminders
Unit EMC2: Practical Assessment
Suggested Answers

Introduction

Course Structure

This textbook has been designed to provide the reader with the core knowledge needed to successfully complete the NEBOSH Environmental Management Certificate. It follows the structure and content of the NEBOSH syllabus.

The NEBOSH Environmental Management Certificate consists of two units of study. When you successfully complete any of the units you will receive a Unit Certificate, but to achieve a complete NEBOSH Environmental Management Certificate qualification you need to pass both units within a five-year period. For more detailed information about how the syllabus is structured, visit the NEBOSH website (www.nebosh.org.uk).

Unit EMC1: Environmental Management	
Element 1	Foundations in Environmental Management
Element 2	Environmental Management Systems
Element 3	Assessing Environmental Aspects and Impacts
Element 4	Planning For and Dealing with Environmental Emergencies
Element 5	Control of Emissions to Air
Element 6	Control of Environmental Noise
Element 7	Control of Contamination of Water Sources
Element 8	Control of Waste and Land Use
Element 9	Sources and Use of Energy and Energy Efficiency
Final Reminders	

Unit EMC2: Practical Assessment
Assessing Environmental Aspects and Associated Impacts - Practical Assessment

Introduction

The Assessments

Open-Book Exam

For everyone studying the NEBOSH EMC1 unit, a new, 'open-book' exam has been introduced to replace the previous invigilated exam. This new format allows candidates to complete their qualification without needing to attend an exam venue, while maintaining the high standards and robust assessment strategy for which NEBOSH is renowned. This study text contains information and advice on how best to prepare for this exam.

NEBOSH has prepared detailed information on the open-book examinations along with a number of useful resources which you can access from the **NEBOSH** website. You must read and familiarise yourself with the resources NEBOSH has provided; RRC's guidance is complementary, not a replacement for NEBOSH guidance.

> **MORE...**
>
> For the guidance that NEBOSH has prepared see *NEBOSH Open Book Examinations: Learner Guide*, available on the NEBOSH website:

An 'open-book' exam means exactly what the name suggests: you are allowed to access your study text and other materials during the exam. However, it will require more than simply copying text to answer the exam questions; you will need to demonstrate that you understand the subject matter and can apply the topics appropriately - and this needs preparing for. Further guidance can be found in the 'Final Reminders' section later.

The NEBOSH Environmental Management Certificate open-book exam differs from the traditional question-and-answer type format and will instead consist of a realistic workplace scenario which may describe a developing situation such as an incident or environmental intervention, and you may be asked to assume a particular role, for example a safety manager. You will then be asked to carry out some tasks and each task will consist of one or more questions, where your answers will be relevant to the scenario.

Another crucial difference between the two exam formats is the duration; since this exam isn't invigilated and requires time to prepare your answers, you have a 24-hour window in which to prepare, complete and submit the exam, therefore allowing for enough time to reflect and detail your answers. It may sound like a lot of time, but there is a lot you will need to do. Don't worry though - we will outline what you need to do in those 24 hours to successfully finish the exam.

NEBOSH specifies 5 hours to complete the exam; the 24 hours allow for you to prepare and choose the best time of day for you to work on the exam - it does not mean you should be working all of that time.

There is a word count currently specified of 3,000 words in total which you would not be expected to exceed. (Note that the word count is subject to change by NEBOSH.)

Be aware though that it will be more difficult to pass the exam if your word count is significantly below the specified figure. The word count might provide a guide as to how extensive your answers should be and will help you to focus on what the exam is asking you to do. If we assume a word count of 3,000 words for example, distributed across all the questions, ten 10-mark questions require around 300 words while 15-mark questions would require around 450 words.

Following the exam there will be a closing interview - you must take part in this for your mark to be awarded.

In addition to RRC's guidance, you must read the *NEBOSH Open Book Examinations: Learner Guide* and the *NEBOSH Open Book Examinations: Technical Learner Guide*, which you can download from the NEBOSH website.

> **MORE...**
>
> NEBOSH has prepared detailed information on the open-book examinations along with a number of useful resources which you can access at: **www.nebosh.org.uk/open-book-examinations/**

Introduction

Practical Application

The practical application tests knowledge and understanding that you have gained from your EMC1 studies. This requires you to assess the environmental aspects and impacts in a workplace.

Further information and help on the practical assessment is given in the Unit EMC2 study text, in the Introduction to Unit EMC2, and in the Unit EMC2 Final Reminders section.

> **HINTS AND TIPS**
>
> As you work your way through this book, always remember to relate your own experiences in the workplace to the topics you study. An appreciation of the practical application and significance of health and safety will help you understand the topics.

Learning Outcomes and Assessment Criteria

Learning outcome The learner will be able to:	Related content	Assessment criteria
Justify environmental management in the workplace using ethical, legal and financial arguments, linking these to wider environmental issues including sustainable development	1.1 – 1.3	Explain the scope and nature of environmental management and key environmental issues Discuss the ethical, legal and financial reasons for maintaining and promoting environmental management Summarise sustainability, its importance, and its relationship with corporate social responsibility
Recognise workplace activities which may be subject to environmental legislation or enforcement	1.4	Understand the influence of international agreements on national environmental laws and standards, and the potential consequences of non-compliance
Understand the requirements of, and work within, an environmental management system, whilst contributing to continual improvement	2.1-2.3	Recognise the key features and appropriate content of an effective EMS based on the requirements of ISO 14001 Discuss the benefits and limitations of introducing a formal EMS into the workplace
Assess environmental aspects and associated impacts, determining significant aspects and evaluating current controls	3.1-3.4	Recognise different types of environmental impact Review and use sources of environmental information Apply the principles and practice of environmental aspect and impact assessment
Support environmental emergency planning	4.1-4.2	Explain the importance of environmental emergency planning Describe suitable emergency preparation and responses
Understand the importance of reducing environmental harm; identify sources of noise, air, and water pollution; and suggest suitable control measures	5-7	Demonstrate awareness of the environmental impacts of noise, air, and water pollution Identify sources of environmental harm and suggest suitable control measures for noise and emissions
Understand the issues associated with waste and support responsible waste management.	8.1 – 8.5	Demonstrate awareness of common waste types, the outlets available for waste, and environmental issues associated with waste and contaminated land Suggest suitable waste management measures, applying the waste hierarchy
Understand the benefits and limitations of a range of energy sources, and suggest suitable measures to increase energy efficiency	9.1 - 9.2	Discuss the benefits and limitations of a range of renewable and non-renewable energy sources
	9.3	Explain how energy efficiency can be increased

These learning outcomes and assessment criteria are published by NEBOSH in the syllabus guide for the course. It is very important that you understand all the learning outcomes.

Introduction

Keeping Yourself Up to Date

The field of environmental health and safety is constantly evolving and, as such, it will be necessary for you to keep up to date with changing legislation and best practice.

RRC International publishes updates to all its course materials via a quarterly e-newsletter (issued in February, May, August and November), which alerts students to key changes in legislation, best practice and other information pertinent to current courses.

Please visit **www.rrc.co.uk/news-resources/newsletters.aspx** to access these updates.

Other Publications

Study Aids

- RRC Environmental Law and Case Law Guide
- NEBOSH Environmental Management Certificate Revision Guide
- RRC Environmental Step Notes

Further Your Studies

- NEBOSH Diploma in Environmental Management – Unit ED1: Controlling Environmental Aspects
- NEBOSH National Diploma in Occupational Safety and Health – Unit ND1: Know – workplace health and safety principles (UK)
- NEBOSH National Diploma in Occupational Safety and Health – Unit ND2: Do – controlling workplace health issues (UK)
- NEBOSH National Diploma in Occupational Safety and Health – Unit ND3: Do – controlling workplace safety issues (UK)

RRC International is continually adding to its range of publications. Visit **www.rrc.co.uk/publications.aspx** for a full range of current titles.

Element 1

Foundations in Environmental Management

Learning Outcomes

- Justify environmental management in the workplace using ethical, legal and financial arguments, linking these to wider environmental issues including sustainable development.
- Recognise workplace activities which may be subject to environmental legislation or enforcement.

Learning Objectives

Once you've read this element, you'll be able to:

1. Explain the scope and nature of environmental management and key environmental issues.

2. Discuss the ethical, legal and financial reasons for maintaining and promoting environmental management.

3. Summarise sustainability, its importance, and its relationship with corporate social responsibility.

4. Understand the influence of international agreements on national environmental laws and standards, and the potential consequences of non-compliance.

Contents

The Scope and Nature of Environmental Management — 1-3
Definition of the Environment — 1-3
The Multidisciplinary Nature of Environmental Management — 1-4
Size of the Environmental Problem — 1-5

The Ethical, Legal and Financial Reasons for Maintaining and Promoting Environmental Management — 1-19
Rights and Expectations of Internal and External Interested Parties (Local Residents, Indigenous People, Supply Chain, Customers and Workers) — 1-20
Outcomes of Incidents — 1-21
The Actions and Implications of Pressure Groups — 1-21
Overview of Compliance Issues — 1-22
The Business Case for Environmental Management — 1-24

Supporting Sustainable Development — 1-28
Definition of Sustainability — 1-28
Importance of Sustainable Development — 1-28
The Business Case for Sustainable Development — 1-31
Role of the United Nations' Sustainable Development Goals (SDGs) — 1-32

The Role of National Governments and International Bodies in Formulating a Framework for the Regulation of Environmental Management — 1-34
International Law — 1-34
The Importance of Knowing and Understanding Local Legislation — 1-39
Meaning of BAT and BPEO — 1-39
The Role of Enforcement Agencies and Consequences of Non-Compliance — 1-39

Summary — 1-41

Exam Skills — 1-42

The Scope and Nature of Environmental Management

IN THIS SECTION...

- Understanding and managing the environment requires knowledge of many topics, including geography, geology, hydrogeology, planning, public health, sociology, pollution and pollution controls.
- Barriers to good environmental management are:
    - The complex nature of the environment.
    - Conflicting demands in an organisation.
    - Difficulties in changing people's behaviour.
- Environmental impacts can be local, national or regional; and international or global.
- Some key environmental issues include:
    - Local pollution from noise, waste, lighting and odour.
    - Carbon emissions and climate change.
    - Loss of biodiversity.
    - Air pollution causing poor air quality.
    - Release of pollutants causing the protective ozone layer in the stratosphere to become depleted.
    - Land grabbing.
    - Use of fossil fuels.
    - Inappropriate disposal of waste.
    - Impacts occurring from poor agricultural practices.

Definition of the Environment

The 'environment' is everything that surrounds us. This encompasses:

- the physical resources of the Earth, including the atmosphere, water, the land and raw materials;
- the living resources of animal and plant life; and
- human populations.

Environmental management is concerned with understanding these elements and how they interrelate. The international environmental management system standard **ISO 14001:2015** defines the environment accordingly as:

> "The surroundings in which an organisation operates, including air, water, land, natural resources, flora, fauna, humans and their interrelationships.
>
> Surroundings can extend from within an organisation to the local, regional and global system."

(Source: **ISO 14001:2015**)

1.1 The Scope and Nature of Environmental Management

> **DEFINITIONS**
>
> **NATURAL RESOURCES**
>
> Land or raw materials that occur naturally in the environment.
>
> **FLORA**
>
> Plant life.
>
> **FAUNA**
>
> Animal life.
>
> **ENVIRONMENTAL MEDIA**
>
> Air, land and water.

The Multidisciplinary Nature of Environmental Management

> **DEFINITIONS**
>
> **GEOLOGY**
>
> Study of the physical materials that make up the Earth.
>
> **HYDROGEOLOGY**
>
> Study of the movement of groundwater in the soil and rocks.
>
> **SOCIOLOGY**
>
> Study of human social activity.

Environmental management clearly has a very broad scope. One of the fascinations of studying the environment is the breadth of topics and disciplines that are involved. Environmental management typically involves concepts from scientific and technical disciplines, e.g. physics, chemistry, biology, geology and engineering, but it also has social and political dimensions, e.g. town and country planning, public health and legislation.

In studying an environmental course such as this, you will be given a general, but not specialist, understanding of a wide variety of topics included in these disciplines.

Barriers to Good Standards of Environmental Management

Many organisations of all types and sizes successfully manage their environmental impacts. But we need to recognise at the outset that the broad scope of environmental management poses a number of barriers to good environmental management, for example:

- **Complexity**

    Organisations are complex, with numerous environmental impacts, such as waste generation and disposal, energy use, emissions to air, or discharges to water. Deciding which impacts to address and how to achieve improvements requires background knowledge of environmental impacts, how they interact, and options for improvement. An understanding of how changes of process or procedure can affect the business is also important.

- **Competing and Conflicting Demands**

    Organisations need to operate in an efficient and effective manner to deliver the right product or service to their customers, on time, and at a competitive price. Commercial companies need to make a profit in order to survive. Individual people in an organisation may be driven by financial, rather than environmental, pressures.

All organisations need to comply with applicable legislation. There is often conflict between environment and health and safety. For example, to protect workers from high dust levels in the workplace, dust is ventilated to the atmosphere. If not controlled, this may cause an environmental problem, with plants being covered in dust, or a nuisance being caused to nearby residents.

- **Behavioural Issues**

    Changing the way people behave in any given situation is one of the most difficult things to achieve. In recent years, the cost of fuel and generally running a car has increased significantly, yet people are still unwilling to give up car ownership and use. This is in spite of significant publicity regarding the negative environmental impact of car use, from contributing to climate change to decreasing local air-quality standards. This is no different from attitudes in the workplace; if we are to be successful in changing behaviour patterns, we must be prepared for it to take time and we must provide people with good reasons to change.

Size of the Environmental Problem

We have seen that the environment covers a wide range of issues and disciplines. When we think about our own organisations, we need to recognise that we can contribute to local, regional and global environmental issues. The following figure illustrates this.

Relationship between local, national and international environmental issues

The various sections of this course cover the main environmental issues in some depth, but let's begin by taking an overview of some of the key local, regional and global environmental concerns that society is dealing with today.

1.1 The Scope and Nature of Environmental Management

Local Effects of Pollution

These can include:

- Poor air quality due to the pollution caused by high levels of vehicle traffic or local industrial processes. For example, vehicles emit a mixture of gases and particulate material that can cause harm to the environment and damage human health. Oxides of sulphur and nitrogen are often referred to as SO_x and NO_x respectively. NO_x can react with atmospheric gases, in the presence of sunlight, to produce harmful low-level ozone.

- Contaminated land from industrial processes where spills or accidents have occurred, leaving ground contaminated with pollutants such as heavy metals (cadmium, lead, etc.).

- Water pollution from accidental spillages from industry.

Pollution by noise, odour and light is becoming an increasing problem and all of these types of pollution are often controlled through legislation. Operating conditions may also be imposed on businesses located in sensitive areas under planning law, or through industrial environmental permits.

High volume of traffic, even at night

Waste is often heavily regulated, e.g. under the **Waste Framework Directive (2008/98/EC)** in the European Union. This is because unregulated and uncontrolled disposal of waste can lead to the spread of disease through contact with the waste itself, or an increase in numbers of vermin species, such as rats, which aid the spread of disease. Waste can also contaminate land and water.

Carbon Emissions and Climate Change

> **TOPIC FOCUS**
>
> **Global Climate Change**
>
> The first decade of the 21st century was the warmest on record and measurements over the last 150 years show that the temperature of the atmosphere has increased by around one degree Celsius. This phenomenon is commonly known as climate change.
>
> There is now strong evidence that climate change is related to pollution of the atmosphere, through the mechanism known as the 'greenhouse effect'.
>
> What is the greenhouse effect? It is actually a natural phenomenon. The Sun irradiates the Earth with energy and as the Earth warms, it emits energy back into space as infrared radiation. Some of this radiation is absorbed by greenhouse gases that occur naturally in the atmosphere (primarily water vapour, carbon dioxide and methane); the effect of this is to reduce heat loss from the Earth. Were it not for the greenhouse effect the temperature of the Earth would be well below zero degrees Celsius.
>
>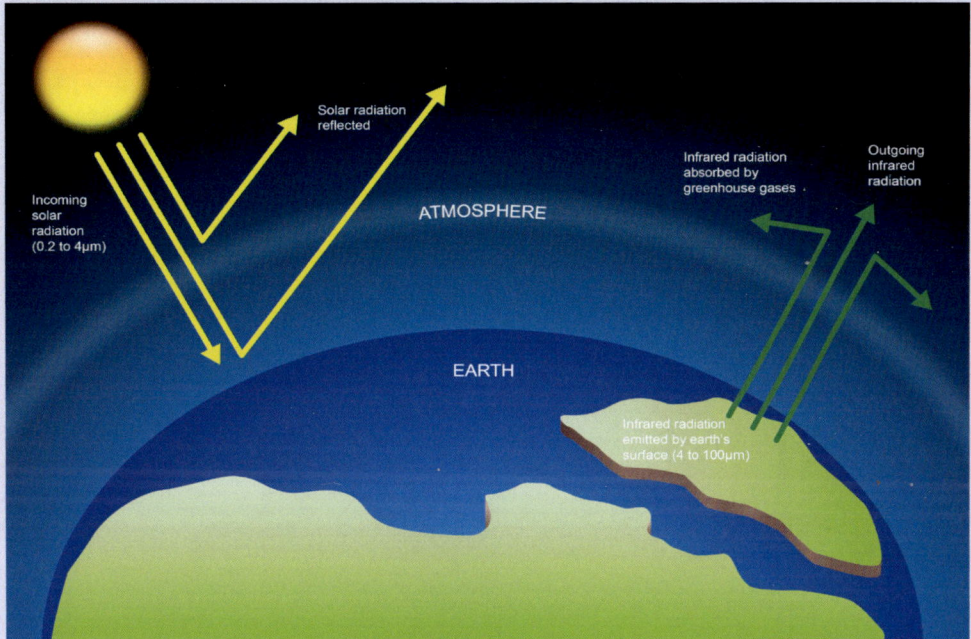
>
> The greenhouse effect
>
> The problem is that burning fossil fuels (e.g. coal, oil, gas, petrol, diesel), which account for more than 85% of the world's energy consumption, releases large quantities of carbon dioxide into the atmosphere. Levels of carbon dioxide in the atmosphere have consequently increased significantly during the past 50 years.
>
> Enhanced levels of carbon dioxide in the atmosphere are now believed to be artificially increasing the greenhouse effect, leading to climate change.
>
> What is so alarming about a warmer planet?
>
> - Sea levels will rise - primarily through the melting of the polar ice caps. This could result in widespread coastal flooding.
> - Climate change - the warming of the Earth is likely to trigger changes in the Earth's climate. This could potentially have very serious consequences. For example, major food-producing areas might begin to suffer droughts, reducing our ability to feed ourselves. There are also likely to be more extreme and disruptive weather events, such as high winds and floods.

1.1 The Scope and Nature of Environmental Management

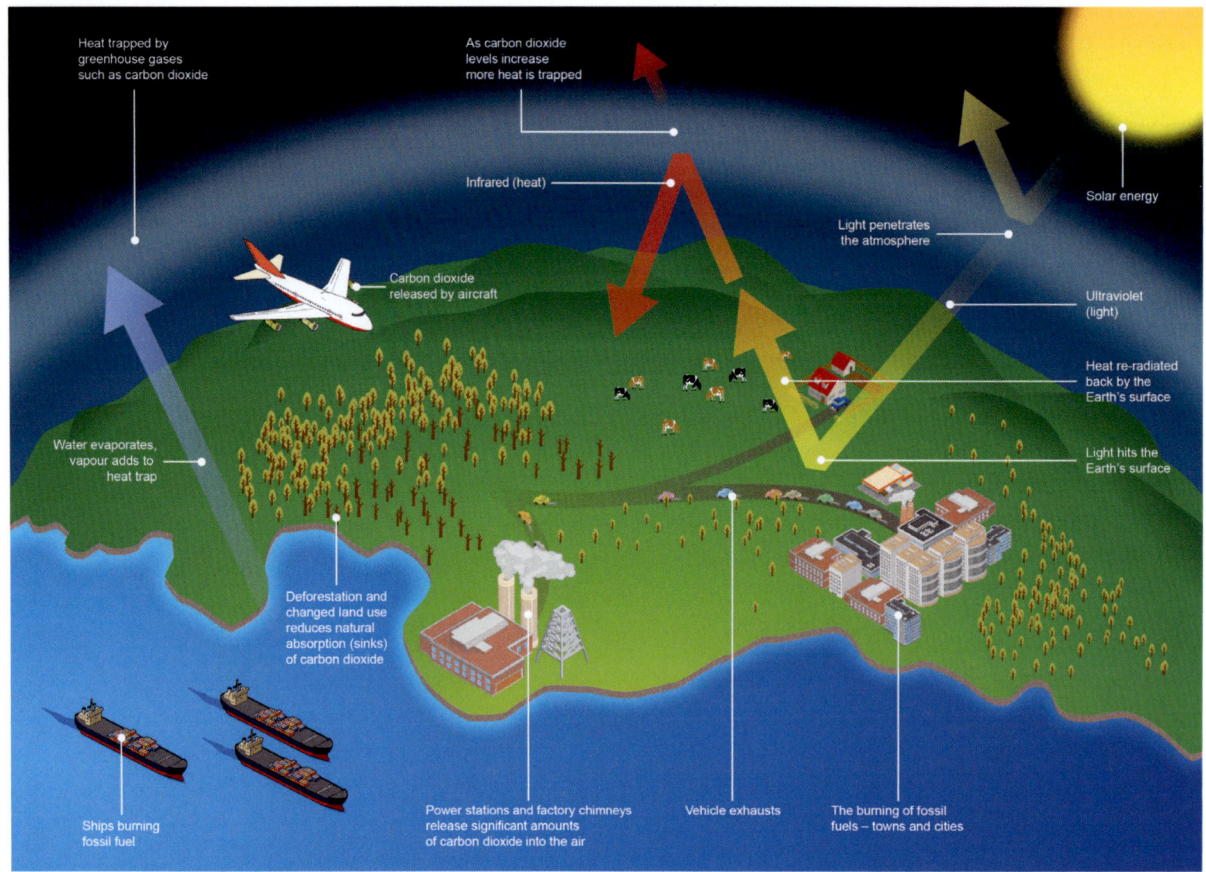

The main man-made sources of carbon dioxide emissions and the greenhouse effect
(reproduced courtesy of Scottish Power)

MORE...

Further information on the science behind climate change can be found in the Intergovernmental Panel on Climate Change (IPCC) *Fifth assessment report - climate change 2013: The physical science basis*, available at:

www.ipcc.ch/report/ar5/wg1

To reduce fossil-fuel burning (mainly power stations and road vehicles), we must:

- **Reduce energy consumption**, e.g. by improved insulation, double-glazing, attention to heating and ventilation, turning lights off.
- **Increase efficiency of energy use**, e.g. through best practice in the operation of plant and processes, use of fuel-efficient vehicles (diesels give about 30% better performance than petrol-driven vehicles but generally emit more air pollutants to atmosphere).
- **Use alternative energy sources**: e.g. wind, water, or nuclear energy.
- **Burn fuels which release less carbon dioxide**: natural gas (methane) produces more than twice as much energy (per kg), and carbon dioxide makes up only 75% of the combustion products compared with coal.

Climate Change Greenhouse Gas Management Hierarchy

The IEMA publication, *Pathways to net zero: using the IEMA GHG management hierarchy*, advocates the use of a hierarchy of control for greenhouse gas emissions. The hierarchy is as follows:

Eliminate
- Influence business decisions/use to prevent GHG emissions across the lifecycle
- Potential exists when organisations change, expand, rationalise or move business
- Transition to new business model, alternative operation or new product/service

Reduce
- Real and relative (per unit) reductions in carbon and energy
- Efficiency in operations, processes, fleet and energy management
- Optimise approaches (e.g. technology and digital as enablers)

Substitute
- Adopt renewables/low carbon technologies (on site, transport, etc.)
- Reduce carbon (GHG) intensity of energy use and of energy purchased
- Purchase inputs and services with lower embodied/embedded emissions

Compensate
- Compensate 'unavoidable' residual emissions (removals, offsets, etc.)
- Investigate land management, value chain, asset sharing, carbon credits
- Support climate action and developing carbon markets (beyond carbon neutral)

GHG management hierarchy

Source: Based on *Pathways to net zero: using the IEMA GHG management hierarchy*, IEMA, 2020

Biodiversity Loss

Biodiversity is simply diversity, or variety, of plants, animals and other living things in a particular area or region. Diversity within the natural environment is important.

The Earth's biological resources are vital to economic and social development because they:

- Provide us with sustainable materials.
- Maintain the quality of our air, soils, waters and climate.
- Contribute to our health and enjoyment of life.

Estimates of global species diversity vary enormously, as it is difficult to estimate how many species there may be in less well-explored habitats, such as untouched rainforests. Rainforest areas that have been sampled have shown a very high level of biodiversity.

Extinction is a fact of life. However, species are now becoming extinct at an alarming rate, almost entirely as a direct result of human activities. Previous mass extinctions evident in the geological record are thought to have been brought about mainly by massive climatic or environmental shifts. Predictions and estimates of future species losses abound. One such estimate calculates that a quarter of all species on Earth are likely to be extinct, or on the way to extinction, within 30 years.

Air Pollution and Ground-Level Ozone

The main causes of air pollutants are vehicle exhaust emissions and industrial activities. Vehicles emit a mixture of gases and particulate material that can cause harm to the environment and damage human health. Oxides of sulphur and nitrogen are often referred to as SO_x and NO_x respectively. NO_x can react with atmospheric gases, in the presence of sunlight, to produce harmful low-level ozone. While we need ozone in the stratosphere part of the atmosphere, at low altitudes where people live it is a poisonous gas. These same SO_x and NO_x gases can also combine with moisture in the atmosphere to produce dilute sulphuric or nitric acid (falling as so-called 'acid rain'), which causes damage to buildings, especially many older buildings that are made from materials such as marble and limestone.

Air Pollution and the Ozone Layer

Life on Earth is protected from the damaging effects of ultraviolet radiation by a layer of ozone molecules (O_3) in the lower stratosphere, between 15 and 25 km above the Earth's surface. Ozone absorbs ultraviolet radiation, one of the major causes of skin cancers. Certain chemicals (ozone depleters) can destroy the ozone layer.

Although ozone has been depleted in other regions, ozone depletion is most dramatic over the polar regions, due to particular upper atmospheric conditions, and a continent-sized hole has developed over Antarctica.

Most ozone depleters are chemically-stable compounds containing the halogen elements chlorine or bromine. These compounds have typically been used as refrigerant gases, as propellants for aerosol sprays, as foam-blowing agents, as solvents and in fire-fighting systems, e.g. chlorofluorocarbons (CFCs), hydrochlorofluorocarbons (HCFCs), carbon tetrachloride, trichloroethane, halons.

These ozone-depleting compounds are very stable and if they are released by human activities they can persist unchanged in the atmosphere until they drift upwards to reach the ozone layer in the stratosphere. At this altitude, the compounds are exposed to higher levels of UV radiation, which liberates charged chlorine and bromine atoms from the parent molecules. These charged atoms are known as 'free radicals' and are highly reactive. Chlorine and bromine free radicals are able to react with, and break down, ozone molecules in a variety of ways, for example:

A chlorine-free radical reacts with ozone to produce chlorine monoxide and molecular oxygen:

$$Cl + O_3 \rightarrow ClO + O_2$$

The chlorine monoxide so formed may then react to break down more ozone:

$$ClO + O_3 \rightarrow Cl + 2O_2$$

International agreements (especially the Montreal Protocol (see later in this element)) are in place to curb the production and use of ozone depleters.

Unfortunately, even if all ozone-depleters were banned today, the chlorine molecules already in the atmosphere would continue to affect stratospheric ozone levels for at least a century.

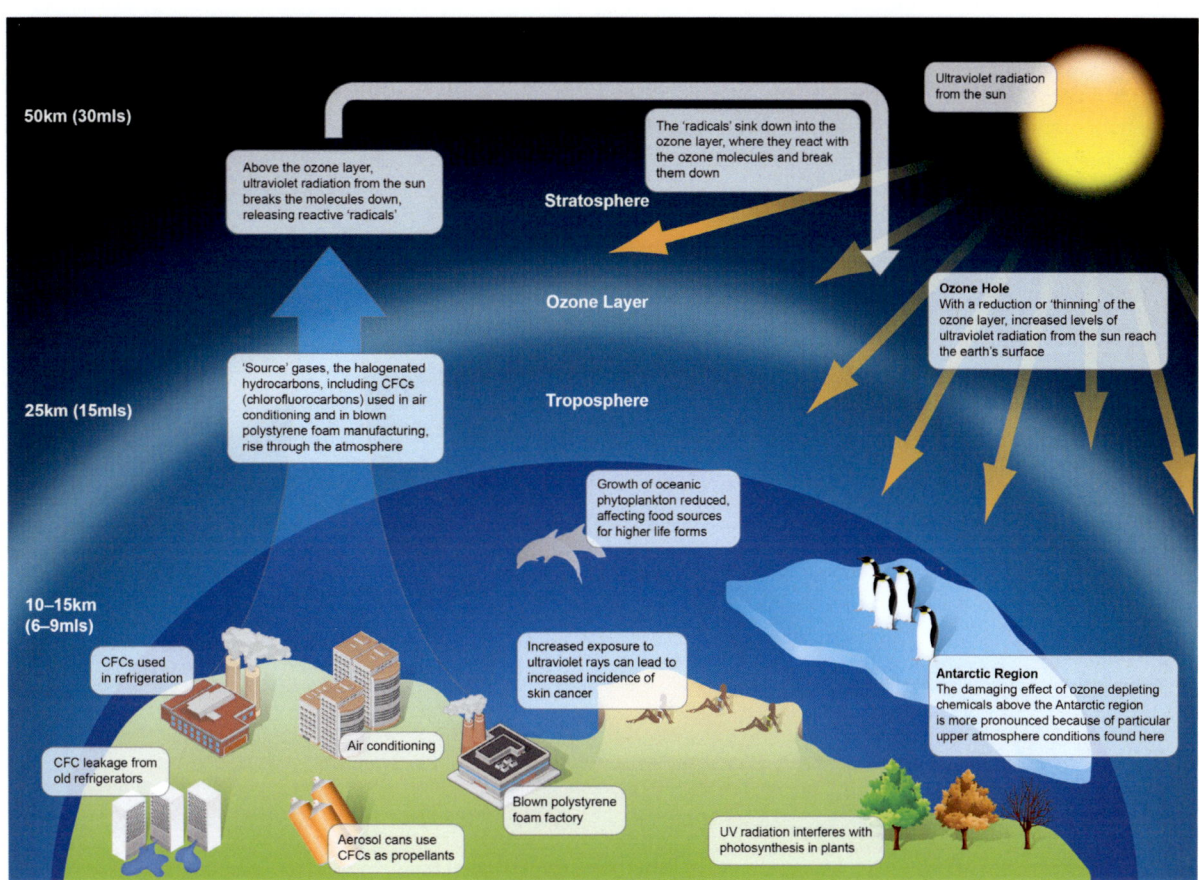

Depletion of the ozone layer

1.1 The Scope and Nature of Environmental Management

Water Resources

Water is an essential resource and is recycled naturally in the environment through the hydrological cycle, as shown in the figure below. The demand for water is increasing due to the increase in population and in the amount of water used by individuals; this is especially the case in developed countries, where water is seen as a plentiful and cheap resource.

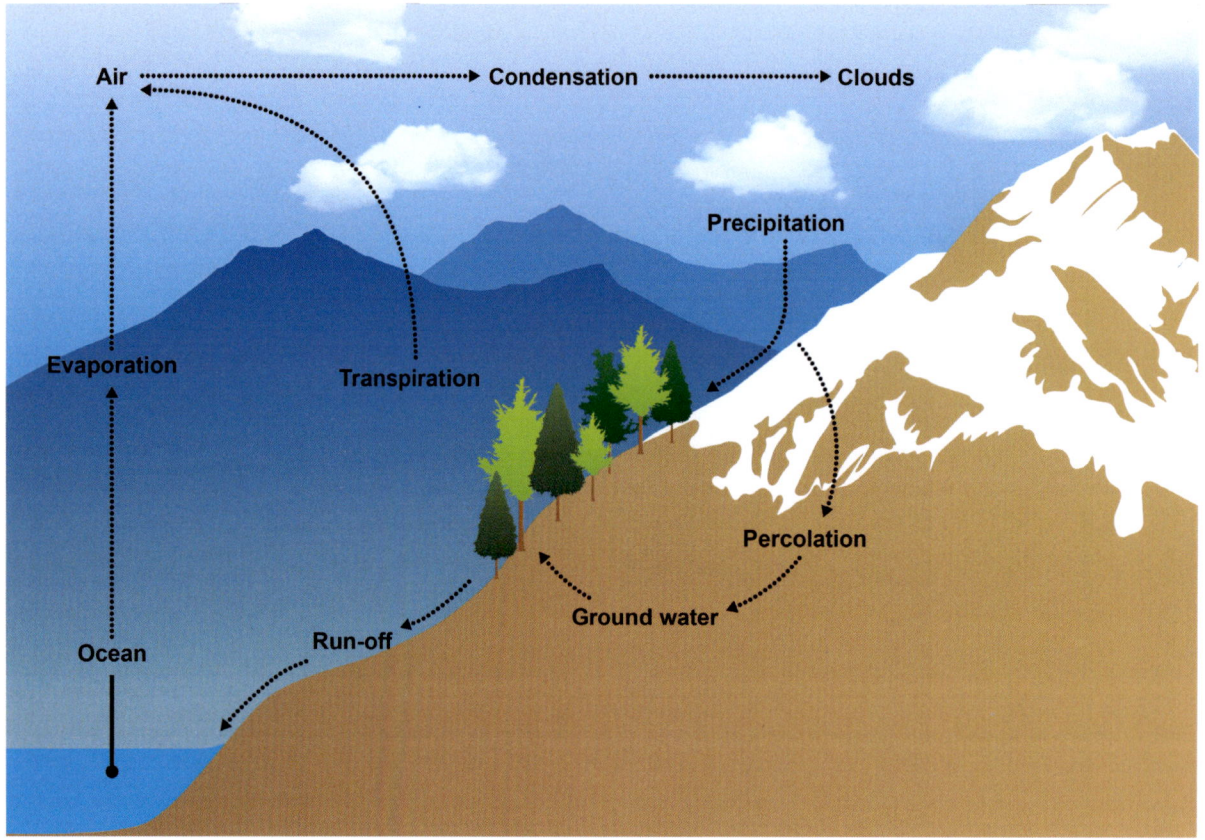

Natural hydrological cycle

Because of this cycle, there can be an accumulation of pollutants through water catchments, making prevention of pollution particularly important.

> **DEFINITION**
>
> **WATER CATCHMENTS**
>
> Areas of land that drain water from rain, snow, etc., into a single water body, such as a river and its tributaries.

Water can be polluted:

- Directly, by discharges to rivers and lakes (point sources).
- Indirectly, through:
  - Run-off from land, particularly contaminated land.
  - Deposition of airborne pollutants (non-point sources) into watercourses.

Water quality can be adversely affected by a number of types of pollutants, such as:

- **Nutrients** - excessive levels of nutrients, e.g. nitrogen and phosphorus, can cause excessive growth of aquatic plants. These, in turn, allow certain species, e.g. green and blue-green algae, to dominate a watercourse, especially in slow-moving sections of rivers, or in lakes. This excessive growth eventually prevents the penetration of sunlight and de-oxygenates the water at night, adversely affecting other plant and animal life. As the source of pollution is usually run-off from a wide area of agricultural land, it is very difficult to control. This process is known as **eutrophication**.

- **Organic wastes** - human and animal effluent, silage and products such as milk are broken down by aerobic (oxygen-using) bacteria. These wastes promote the growth of aerobic bacteria and, because they have a high oxygen demand, they reduce the level of oxygen in the water that is available to higher life forms, such as fish. These pollutants are said to have a high **Biological Oxygen Demand (BOD)**.

- **Immiscible liquids** - these are liquids that do not mix with water, such as oil-based products. They often form a layer on the top surface of the water and prevent the transfer of oxygen and other gases between the water and the atmosphere. They often also coat the leaves of plants and are toxic to animals.

- **Sedimentation** - suspended solids are a very common type of pollutant and can be caused by many processes, both natural and man-made. A natural process could be run-off from a hillside, while a man-made cause could be dewatering a quarry, or washing down a sand and gravel processing yard. The key impacts of excess sediment in water are:

    - Reduced penetration of sunlight into the water which therefore reduces the ability of aquatic plants to photosynthesise, thereby reducing oxygen levels in the water.

    - Physical problems such as damaging plants and destroying breeding grounds for aquatic animals.

- **Flow-rate changes** - plants and smaller aquatic animal life are adapted to specific conditions, including flow rate. Any significant changes to these conditions can result in the reduction or changing of the plant and animal species able to survive in that environment.

- **Pathogens** - disease-causing organisms such as bacteria, viruses and parasitic worms can enter the water in untreated sewage and animal wastes.

- **Temperature changes** - warm water can hold less oxygen than cold water. Discharging warm water into cold water, often from an industrial process, can cause oxygen levels to drop and so have an adverse impact on the potential for aquatic plants and animals to thrive in that environment. This problem is worse in slow-moving waters, as the oxygen levels become further depreciated over time and are not replenished by fresh inflowing water.

- **Harmful chemicals** - such as methyl mercury, lead, and salts may be harmful to aquatic plant and animal life. Some of these may be taken up into the food chain and become harmful to humans.

- **Acidity** - pH values of acid rain as low as 2.1 have been measured - the equivalent of raining vinegar! The rain ends up in watercourses. Adverse effects of acidity include:

    - Fish dying in Scandinavian lakes.

    - Distorted, diseased and dying trees in European forests.

    - Damage to buildings, metals, rubber, plastic and nylon.

    - Aluminium and heavy metals such as lead, tin, copper and manganese present in soils being leached out. Metals are toxic to plant and animal species.

1.1 The Scope and Nature of Environmental Management

> **DEFINITIONS**
>
> **EUTROPHICATION**
>
> Excessive plant growth in water caused by the addition of nutrients and resulting in a depletion of oxygen and water quality in the watercourse.
>
> **ORGANIC**
>
> Class of chemical compounds based on carbon - usually includes materials of, or derived from, animals and plants.
>
> **BIOLOGICAL OXYGEN DEMAND**
>
> A laboratory procedure for determining the amount of oxygen needed by organisms to break down organic materials in water.
>
> **PHOTOSYNTHESIS**
>
> The process that plants and algae use to convert sunlight and carbon dioxide into energy.
>
> **FOOD CHAIN**
>
> Represents the 'food' relationship between organisms and species in an ecosystem. For example, phytoplankton (tiny marine, plant-like organisms, which manufacture their own food using light) are eaten by fish, which are eaten by larger fish, which are eaten by humans.
>
> **FLOW RATE**
>
> Volume of water passing through a river over a set period of time.

Ocean Pollution

Significant issues include:

- Discharge of nitrates and phosphates causing oxygen depletion.
- Plastic pollution.
- Ocean acidification as a result of carbon dioxide absorption.

We will cover these in more detail in a later element.

Deforestation, Soil Erosion and Land Quality

> **DEFINITIONS**
>
> **BIOSPHERE**
>
> The part of the Earth and its atmosphere in which living things are found.
>
> **ECOSYSTEM**
>
> Refers to a community of interrelated species in a defined area.

The world's forests have major influences on the biosphere. Deforestation is the removal of trees from large areas of land, e.g. the Amazon basin in South America, and can lead to a number of negative environmental impacts.

- Deforestation can contribute to climate change in the following ways:
  - Burning and decay of wood releases carbon dioxide into the atmosphere.

- Trees and other plants photosynthesise - this involves removing carbon dioxide from the atmosphere to produce oxygen, thereby reducing atmospheric carbon dioxide levels. If large forests are removed, less carbon dioxide is removed from the atmosphere.
- The water cycle can be significantly affected. Trees take groundwater through roots, which is emitted into the atmosphere. When deforestation occurs, the lack of trees and other plants means that water is not evaporated and local climates are much drier.

- The cohesion of the soil is reduced by deforestation, resulting in:
  - Fertile agricultural soils being eroded.
  - Increased risk of landslides on steep slopes.
- A reduction in forest cover means that surface water run-off will increase, which may result in flash floods and increase the risk of localised floods compared with what would occur if the forest cover were present.
- Deforestation can result in a decrease in biodiversity. This can lead to a reduction in genetic variation. Genetic variation can lead to many agricultural benefits, such as development of crops that are resistant to pests, or have the ability to grow in poor-quality soils.
- Forests often contain many plants that are yet to be discovered, some of which may have properties that can be used to fight disease and ill health.

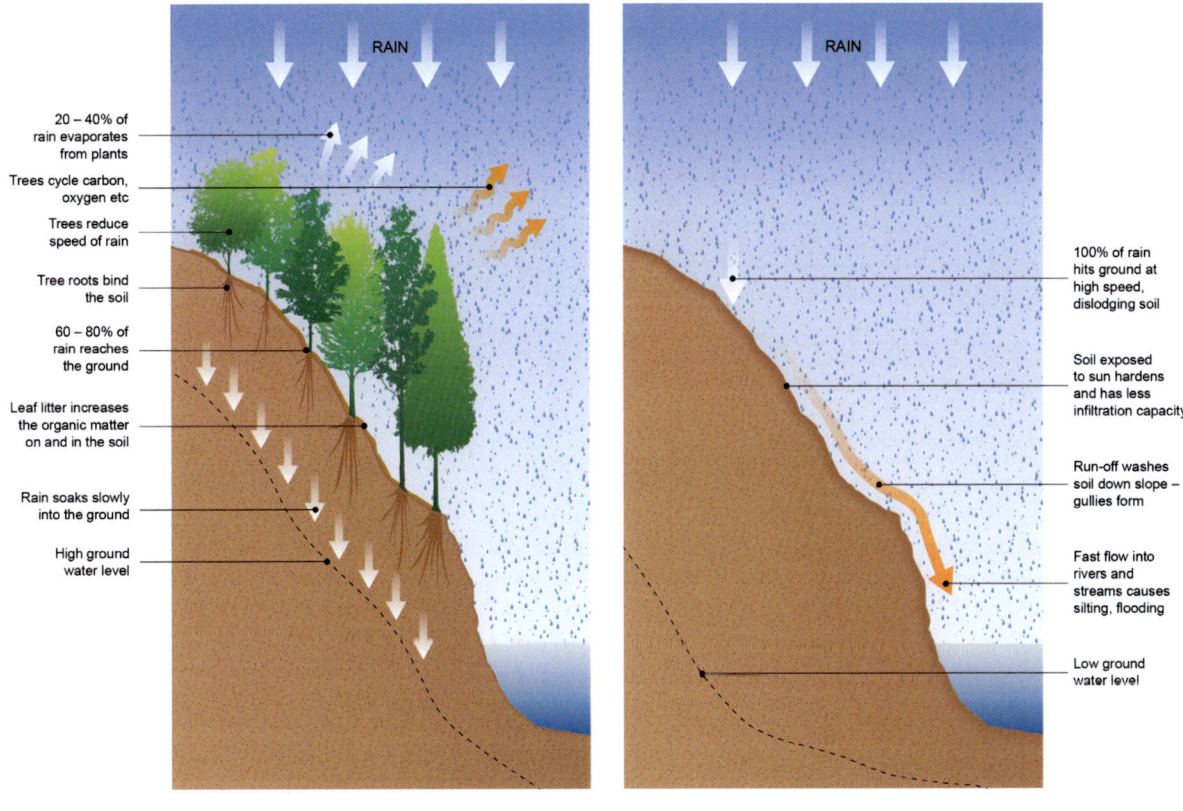

Some effects of deforestation

1.1 The Scope and Nature of Environmental Management

Although short-term economic gains can be made from converting forested areas to agricultural land, or overly exploiting forests for wood products, if forests are not managed sustainably in the long term, deforestation will lead to a loss of long-term income.

Shifting cultivation ('slash and burn' agriculture) disrupts the forest ecosystem, particularly when it is on a large scale. Trees are cut from the soil, burnt and the ash returned. Any nutrients are quickly leached away as they are contained only in the top few centimetres of the soil in tropical forests; therefore, the cleared plots soon become infertile.

Material Resources and Land Despoliation

Access to land is vital to ensure the availability of food and to secure the livelihoods of poor rural populations. Land grabbing involves taking land for another use, such as intensive agricultural practices like palm oil production. This means that indigenous peoples, nomads, etc. have no (or reduced) access to food and livelihood security. Such activities will also capture water resources, meaning that people have no access to drinking water.

Deforestation

Land grabbing is often a non-sustainable model of land use, which:

- Causes forced evictions.
- May seriously affect the environment, e.g. changes in land use, use of pesticides, excessive water use and reduction of biodiversity.
- Substantially reduces natural resources.

Inequal Distribution of Impacts and the Supply Chain

> **DEFINITION**
>
> **SUPPLY CHAIN**
>
> A chain of people, activities, operations and organisations that bring a product or service to the customer. It could involve, for example, those who: extract raw materials; transport the materials to manufacturers; manufacture the component parts; manufacture the final product; transport the final product to retailers; retail the final product to the customer.

As a result of globalisation, product manufacture is often concentrated in one specific region and the resulting product is shipped to other countries. As a result, the impacts are not evenly distributed as the country where the product is consumed is not having to cope with the impacts of its manufacture. Some organisations have limited environmental impacts but their supply chain impacts, such as how their products are grown or manufactured, can be significant.

Energy Supplies

Fossil fuels are attractive as an energy source. Being highly concentrated, they allow significant amounts of energy to be trapped in relatively small volumes. They also allow for easy distribution.

However, the adverse environmental impacts resulting from the combustion of fossil fuels are significant. They include:

- **Acid Rain**

    Acid gases resulting from fossil-fuel combustion combine with water vapour to create acid rain, which corrodes buildings, damages and kills trees and destroys life in rivers and lakes.

- **Smog**

    When gases from vehicle exhausts react with sunlight, smog is formed over cities, damaging trees and crops and affecting health.

The Scope and Nature of Environmental Management 1.1

- **Dwindling Resources**

    Fossil fuels cannot be rapidly reproduced. It takes millions of years to produce coal, gas and oil, and existing reserves will eventually run out.

- **Health and Welfare**

    Energy production from fossil fuels can have significant effects on health. Emissions from transportation include acute effects on certain people, e.g. streaming eyes, coughing, breathing difficulties and asthma attacks. Smog can irritate the lungs, cause bronchitis and pneumonia, and decrease resistance to respiratory infections.

- **Deforestation**

    Every year, it is estimated that an area of tropical rainforest one and a half times the size of England is lost around the world. This removes a carbon sink and releases carbon dioxide into the air.

- **Climate Change**

    A number of greenhouse gases are emitted during the combustion of fossil fuels, the most significant being carbon dioxide.

- **Thermal Pollution**

    Heated water from fossil-fuel power production can have effects on the aquatic systems to which it is discharged.

Innovations in Food and Fuel

Innovative practices across the supply chain could significantly increase the sustainability of food and fuel production and distribution. Changes across the supply chain are depicted below:

Source: Farming First, *Innovations for Sustainable Food Systems* (https://farmingfirst.org/food-systems#home)

MORE...

https://farmingfirst.org/food-systems#home

Waste Disposal and the International Waste Trade

The inappropriate storage, treatment, transport and disposal of waste can have numerous environmental impacts, including:

- Land contamination.
- Water pollution.
- Air pollution.
- Exposure of people to harmful substances.

Historically, numerous serious issues have occurred where waste has been exported from a developed country to one that has no facilities to treat or dispose of the waste in a manner that does not impact significantly on the environment and/or human health.

An international law has therefore been developed to control the international waste trade. The Basel Convention on the Control of Transboundary Movements of Hazardous Wastes and their Disposal requires that exporters of hazardous waste receive written consent from importing nations prior to the waste being transported. An amendment in 2019 added certain types of mixed and contaminated plastic waste to the Convention. Various developing countries have gone further and banned imports of waste through their own policies or legislation.

> **MORE...**
>
> More information on the Basel Convention is available at:
>
> www.basel.int

Agricultural Issues Arising from Global Trade

As a result of globalisation, poor environmental practices can result from food and other products being grown in a country for export to other countries (commonly trade between developing and developed countries). Such issues include:

- Clearing of forest and swamplands for cash crops.
- Using water for irrigating crops or watering animals that would be otherwise used by the local population.
- Use of harmful pesticides that may affect biodiversity and the health of the local population.
- Soil erosion, leading to loss of fertile soil and reduction in downstream water quality.
- Fertiliser run-off, leading to poor water quality.
- Creation of poorly managed landfill facilities at agricultural sites.

One example where such trade creates many environmental and other issues is the global trade in palm oil. Palm oil is produced from the oil palm tree and is grown in countries in South East Asia, Central and West Africa and Central America. It is exported for use in a wide range of products, such as foodstuffs, personal-care products and biofuel. However, it has some significant impacts on the environment, such as driving deforestation, being a key cause of the demise of critically endangered species (such as the Sumatran orangutan and Asian rhinoceros) and destroying cash crops that are owned by indigenous people.

> **STUDY QUESTIONS**
>
> 1. List the three media that make up the 'environment'.
> 2. Briefly explain the terms 'greenhouse effect' and 'global warming'.
>
> (Suggested Answers are at the end.)

The Ethical, Legal and Financial Reasons for Maintaining and Promoting Environmental Management

IN THIS SECTION...

- The three main reasons why an organisation has to manage environmental impacts are: ethical (or moral); legal; and financial (or economic).
- Key stakeholders for environmental management are local residents (including indigenous people), the supply chain, customers and workers.
- Pressure groups are organisations whose aim is to influence governments or businesses at local, national and international levels.
- Significant direct and indirect costs can occur from poor environmental management.

> **TOPIC FOCUS**
>
> **General Benefits of Good Environmental Management**
>
> An organisation can benefit in many ways if it maintains a high level of environmental performance:
>
> - Better relations with communities local to an organisation by participation in local environmental schemes.
> - Minimised energy costs.
> - Decreased cost for managing wastes.
> - Improved corporate image resulting in many business benefits.
> - The organisation may be competitive on an international basis if it implements an environmental management system to internationally recognised standards, such as **ISO 14001**.
> - Improved sales due to enhanced environmental performance of products or services.
> - Opportunities for innovation, including improving existing products or developing new products.
> - Reduction in the chances of an incident occurring that could cause significant environmental impacts.
> - Provides for a better legal defence should an incident occur.
> - Reduced insurance premiums.
> - Improved access to finance, such as grants, loans and investments.
> - Product with a minimal impact on the environment may stand out from other products.
> - Improved staff recruitment to a reputable company that understands its environmental responsibilities.
> - Reduced chance of incidents occurring leading to prosecution for breaches of environmental law.
> - Reduced abatement control costs.

Rights and Expectations of Internal and External Interested Parties (Local Residents, Indigenous People, Supply Chain, Customers and Workers)

Morally at least, people have a right to enjoy the environment around them. People (including workers) justifiably expect businesses (which may have a large pollution potential) to exercise reasonable care to ensure that they don't pollute the environment. If an environmental 'incident' occurs (such as unreasonable noise in a residential area, unpleasant odour, littering, water pollution, etc.), the expectation is that the person responsible should discontinue, remediate and perhaps also compensate.

> **DEFINITIONS**
>
> **REMEDIATE**
>
> Repair the harm done to the environment by returning it to its unpolluted state. This could involve removing pollutants and could extend as far as restocking waterways with species affected by a pollution incident, or replanting damaged vegetation.
>
> **SUPPLY CHAIN**
>
> A chain of people, activities, operations and organisations that bring a product or service to the customer. It could involve, for example, those who: extract raw materials; transport the materials to manufacturers; manufacture the component parts; manufacture the final product; transport the final product to retailers; retail the final product to the customer.

There is an expectation that:

- People should have a say in any proposed scheme that is likely to seriously affect their local environment (some of which will be covered under town and country planning legislation).
- Information on such matters should be accessible. (In relation to environmental information held by public bodies, specific laws exist in some parts of the world.)
- Environmental information should be transmitted down the supply chain (so that people are informed about things such as environmental hazards, or responsible disposal of waste, and can take appropriate action).

There are many environmental pressure groups, customers and others who are quite prepared to boycott businesses for perceived irresponsible and unethical behaviour toward the environment. Workers will also influence an organisation to improve environmental performance.

It is estimated that there are around 370 million indigenous people in approximately 70 countries around the world. Indigenous people are those that existed before the arrival of people with different cultural or ethnic origins. Their economic, cultural and political practices are distinct from those of the main society in which they exist. Such groups around the world include the Mayas in Guatemala, Aborigines in Australia and the Maori in New Zealand.

Indigenous people in some areas of the world can be a neglected part of society, experiencing problems such as discrimination and lack of access to political representation, participation, or social services. As with any cultural group, they aim for recognition of their identity, ways of life and rights to land and natural resources.

1.2 The Ethical, Legal and Financial Reasons for Maintaining and Promoting Environmental Management

Outcomes of Incidents

Environmental incidents can have serious and far-reaching consequences, resulting in environmental and human harm and legal and economic effects on organisations. Below are examples of some major incidents resulting in such outcomes.

Environmental and Human Harm

Location and Year	Incident	Outcome
Minamata Bay, Japan, 1953-1960	The Chisso Corporation's factory discharged methyl mercury in its wastewater into the bay over a number of years.	The methyl mercury built up to high levels in fish. When contaminated fish were eaten by humans, it caused chronic mercury poisoning affecting the central nervous system, sensory impairment, numbness, dizziness, loss of vision and hearing, coma and, in some cases, death. It is thought to have affected more than 3,000 people.
Bhopal, India, 1984	A leak of methyl isocyanate gas from the Union Carbide factory.	This killed 2,000 people and affected many more. Acute effects included burning in the eyes and respiratory tract, breathlessness, stomach pains, vomiting and choking, and pulmonary oedema. Many deaths resulted from choking. Many more people are thought to have died from the long-term consequences of exposure to the gas.
Chernobyl, Ukraine, 1986	Overheating of a water-cooled reactor caused an explosion and the release of radiation from a nuclear power plant.	30 people were killed immediately. The radioactive particles spread across Scandinavia and Western Europe. Several thousand people could still die owing to the effects of the radiation.
Basel, Switzerland, 1986	A fire at a chemicals factory resulted in fire water carrying mercury and pesticides into the river Rhine.	Half a million fish were killed and drinking water was contaminated and unusable.
Buncefield, UK, 2005	A leak of petrol from an oil storage depot resulted in an explosion.	Much of the site and buildings in the vicinity were seriously damaged or completely destroyed. Drinking-water sources in the area were contaminated by fire-water run-off. No lives were lost.
Gulf of Mexico, 2010	Explosion on the Deepwater Horizon oil well resulted in the spillage of 23 million litres of oil into the Gulf of Mexico.	More than 160 kilometres of the USA's Louisiana coastline were affected and 11 workers on the rig died. More than 155,000 square kilometres were closed to fishing. Long-term economic and ecological effects are yet to be assessed.

The Actions and Implications of Pressure Groups

Pressure groups are organisations whose aim is to influence governments or businesses at the local, national and international levels. They can cover a single issue (and, as such, are often classed as 'cause'-based), or may cover multiple issues. Such groups change over time and they can be involved in numerous activities.

International environmental pressure groups include Greenpeace, Friends of the Earth, the International Union for Conservation of Nature (IUCN) and the World Wide Fund for Nature (WWF).

> **MORE...**
>
> Greenpeace:
> www.greenpeace.org/international
>
> Friends of the Earth:
> www.foei.org
>
> International Union for Conservation of Nature (IUCN):
> www.iucn.org
>
> World Wide Fund for Nature (WWF):
> www.wwf.org.uk

Some of the key ways in which pressure groups exert influence include:

- Lobbying - discussing concerns with decision-makers, such as those involved in making law.
- Direct action - for example, Greenpeace influenced Shell Oil's attempt to dump the Brent Spar platform in the North Atlantic through occupation of the platform as it was being towed to its dumping point, and also staged protests at service stations.
- Publicity - pressure groups often try to generate as much publicity as possible for the issue in question in order to gain positive media attention.
- Legal action - pressure groups may fight their cause by legal means, inquiring about the legality of the issue.

The implications of pressure groups can be significant for organisations, as they may invoke the following responses:

- Reduced sales.
- Raising consumer awareness of an issue.
- Increasing the costs of a business through improved risk controls, etc.
- Changing current business practices.
- Influencing the making of law and government policy.
- Damaging the reputation of an organisation or product/service.

Overview of Compliance Issues

The law is a collection of rules designed to regulate and control the conduct of citizens, laid down by those in authority and enforced by its officials.

There is a strong body of international law that covers environmental issues. Environmental issues such as global climate change and ozone depletion do not recognise countries' borders. Often, countries will also develop their own laws that are based on compliance with international law, or include legal controls on other environmental issues. Groupings of sovereign countries, such as the European Union, may also be another influence on the laws that a country makes.

The Ethical, Legal and Financial Reasons for Maintaining and Promoting Environmental Management

Legal Rights of Individuals

The Universal Declaration of Human Rights was adopted by the United Nations General Assembly in 1948. Among other human rights it sets out the key legal rights of individuals:

- **Article 6**

    "Everyone has the right to recognition everywhere as a person before the law."

- **Article 7**

    "All are equal before the law and are entitled without any discrimination to equal protection of the law. All are entitled to equal protection against any discrimination in violation of this Declaration and against any incitement to such discrimination."

- **Article 8**

    "Everyone has the right to an effective remedy by the competent national tribunals for acts violating the fundamental rights granted him by the constitution or by law."

- **Article 9**

    "No one shall be subjected to arbitrary arrest, detention or exile."

- **Article 10**

    "Everyone is entitled in full equality to a fair and public hearing by an independent and impartial tribunal, in the determination of his rights and obligations and of any criminal charge against him."

- **Article 11**

    (1) "Everyone charged with a penal offence has the right to be presumed innocent until proved guilty according to law in a public trial at which he has had all the guarantees necessary for his defence.

    (2) No one shall be held guilty of any penal offence on account of any act or omission which did not constitute a penal offence, under national or international law, at the time when it was committed. Nor shall a heavier penalty be imposed than the one that was applicable at the time the penal offence was committed."

Guidance and Codes of Practice

A large volume of policy, in the form of official guidance notes, is issued by government departments and regulatory agencies for the guidance of public decision-making bodies. This is used to fill out the details of each pollution regulatory regime. These set out exactly how the requirements of the law may be complied with.

Voluntary Requirements

Although not law, voluntary environmental requirements can sometimes be just as important in influencing organisational behaviour.

There are many voluntary requirements that may be important such as agreements with non-governmental organisations, agreements with public authorities and voluntary codes of practice (often produced by trade bodies). Others include:

- Environmental labelling schemes - such as the Forest Stewardship Council (FSC) scheme and Fairtrade.
- International standards - for example development of an environmental management system to the ISO 14001 standard.
- Environmental performance guidelines - such as the International Finance Corporation (IFC) Guidelines and those produced by the European Bank for Reconstruction and Development (EBRD).

Environmental Reporting

Corporate Environmental Reporting (CER) aims to communicate an organisation's environmental performance to its stakeholders. In most countries, including the UK, it is voluntary, but in some countries such as New Zealand, Denmark and the Netherlands it is mandatory.

Specific types of environmental reports are sometimes required by law. In the UK, for example, large organisations are required to report their energy usage and carbon emissions.

Management system standards also sometimes require public environmental reporting such as the Eco Management and Audit Scheme (EMAS) that we will cover later.

Environmental information often forms part of a Corporate Social Responsibility (CSR) report or a sustainability report that covers social and economic issues as well as environmental issues. Guidelines have been set up with regard to the development of sustainability reports such as those produced by the Global Reporting Initiative (GRI).

The Business Case for Environmental Management

It is possible to set out three main reasons for a business to actively and positively manage its environmental impacts:

- Moral or ethical.
- Legal.
- Financial.

These reasons are often not entirely separate, in that both moral and legal reasons result in a financial impact. However, it is worthwhile attempting to separate them to some extent for the purposes of examining the direct drivers behind them.

Moral

The moral reasons for improving environmental performance are a mixture of social and ethical influences and can be summed up as 'doing what is right'. Every individual and every organisation will have a set of beliefs about what is right and how they believe they should go about doing their business. These beliefs will have been developed over time owing to a mix of social culture and pressures from others, such as the public and the media.

There are a number of stakeholders who play a significant part in developing the moral environment in which a business may function. Underlying these beliefs is the principle of sustainable development (discussed further below). It is increasingly important for businesses to take account of a wider range of stakeholders than may previously have been considered. This group now includes:

- **Consumers** - are now more aware of both their impact on the environment and their ability to bring about change in large organisations when they act together.
- **Local communities** - are no longer willing to suffer pollution and other negative environmental or health effects for the sake of the potential increased prosperity a business may bring to an area.
- **Employees** - like working for an organisation with good environmental performance, and some organisations may use this as a factor in aiding recruitment and retention.
- **Insurance companies** - are likely to reward organisations with a good record in environmental management and which can show evidence of positive management of their impacts.

Legal

The repercussions of failing to comply with relevant legislation can be severe for both companies and individuals. In the most serious cases, fines may be unlimited and individuals may be sent to prison. In the UK, for example, this may be for up to five years for some offences.

Financial

Aside from the indirect financial impacts that may arise through the moral and legal routes, there are also direct financial benefits from a good standard of environmental performance.

The 'Polluter Pays' principle that has been developed through the European Union provides a range of financial tools that can be used to encourage or force organisations to account for the pollution they create through the balance sheet. Some examples of these charges are:

- The **Climate Change Levy** applied to commercial fuel sources, such as coal, gas, electricity and Liquefied Petroleum Gas (LPG) - some of this income is used to fund organisations such as the Carbon Trust, which works with individuals and organisations to reduce their carbon emissions
- The **Landfill Tax** applied to all waste deposited in landfill to encourage a movement away from the use of landfill as a means of disposal.
- The **Aggregates Tax** applied to virgin aggregates used in construction - this is designed to encourage the re-use and recycling of material in preference to the continued use of virgin material.
- **Charges for Environmental Permits, etc.**, based partly on the level of pollution and partly on the standard of environmental performance.

Direct vs Indirect Costs

This is an often-used comparison for costs associated with health and safety. It is really the flip side of the business case for managing risks. Environmental costs are no different in this respect. These costs may be divided into two parts:

- **Direct Costs**

    These are the calculable costs arising from an accident and/or any claim for liability in the civil or criminal courts. They include:

    - repairs or replacement of damaged equipment and buildings;
    - remediation;
    - product loss or damage;
    - loss of production;
    - public and/or product liability;
    - fines;
    - legal fees; and
    - increases in insurance premiums.

- **Indirect Costs**

    These are costs that may arise as a consequence of the event, but do not generally actually involve the payment of money. They are often largely unknown, but it is estimated that, in certain circumstances, they may be extremely high. They include:

    - business interruption;
    - loss of orders;
    - cost of time spent on investigations; and
    - loss of corporate image.

Insured vs Uninsured Costs

Employers will invariably take out insurance to cover themselves against potential losses caused by such events as fire and theft. In certain parts of the world they are also required by law to have insurance against certain types of liability (e.g. employers' liability insurance). However, many of the costs involved in respect of accidents are not covered by insurance.

Many of the costs involved in respect of accidents are not covered by insurance

1.2 The Ethical, Legal and Financial Reasons for Maintaining and Promoting Environmental Management

Insurance varies, but there are often many exclusions in the small print. Uninsured costs usually include all indirect costs, as well as those relating to loss of production as a result of many types of incident. In addition, the insurance to cover loss in respect of certain events may be void where it can be shown that the employer has not taken adequate precautions to prevent the incident. Uninsured losses can be many times greater than insured losses.

Non-compliance with law can lead to a number of sanctions, such as fines and, for more serious offences, imprisonment. In the UK, for example, non-compliance with the Environmental Protection Act 1990 (a key environmental law) can lead to an unlimited fine and/or two years in prison. Non-compliance may also lead to the polluter having to pay compensation to the person who has been wronged in relation to an environmental incident. Examples of compensation that might have to be paid include payment of the costs incurred for cleaning up an oil spill, or restocking a river with fish.

Legal and Economic Effects on Organisations

There are a number of legal and economic effects on organisations responsible for environmental incidents:

- **Cost of Fines following Prosecution**

    Total, BP and Shell were fined a total of £5.35 million (around $8.3 million) for their involvement in the Buncefield oil-depot explosion. Another £4 million ($6.3 million) was awarded in costs.

- **Clean-Up Costs**

    These can extend to millions of pounds. Following the Deepwater Horizon oil-well explosion in the Gulf of Mexico, BP were reported to have spent $1 million just on containment of the spill. Clean-up costs are thought to amount to over $14 billion.

- **Compensation Payments**

    In 2010, compensation payments for the losses suffered in the 2005 Buncefield oil-depot explosion had still not been settled. Much business disruption was experienced and individuals lost their employment and homes. Total was liable for compensation payments of £750 million ($956 million).

- **Indirect Costs Resulting from Loss of Credibility and Support in the Market**
    - BP's share price plummeted following the explosion in the Gulf of Mexico.
    - It is difficult to estimate the loss to a company where customers choose to go elsewhere. Smaller companies are likely to face bankruptcy.

Different Levels of Standards

The top tier of law in many countries is that of international environmental law. International law can take many forms, including the following:

- **Treaty** - this term is used for matters of high importance that require a solemn agreement.

- **Convention** - describes a multilateral agreement with numerous parties. Conventions are usually open for participation by many nations, or the full international community.

- **Protocol** - generally, this is an agreement that is less formal than agreements using the terms 'Treaty' or 'Convention'. One type of Protocol provides more detailed implementation of the general requirements of a Convention. For example, the Montreal Protocol on Substances that Deplete the Ozone Layer 1987 provides further implementation of the Vienna Convention for the Protection of the Ozone Layer 1985.

- **Declaration** - this is a term used for numerous, usually non-legally-binding, agreements made where the parties do not want to create a legally-binding agreement but do want to declare aspirations, e.g. the Rio Declaration on Environment and Development 1992.

A good understanding of local legislation is important to effectively manage environmental issues

The Ethical, Legal and Financial Reasons for Maintaining and Promoting Environmental Management

Adoption is the term used when an international agreement is developed and established. This is achieved by the consent of states that have participated in development of the agreement, usually by a vote. **Ratification** is the formal act whereby a state is bound by the requirements of an agreement. The time prior to ratification gives states a time period to gain approval of the agreement at the national level. **Entry into force** is the date by which states are bound by the requirements of the agreement.

The next level of law is that of a group of nations, often in a regional area. An example is the European Union. Types of European Law include:

- **Regulation** - applies directly to the intended target (normally member states). There is no requirement to assimilate into national laws.
- **Directive** - binding on EU member states with respect to the objectives to be achieved, but the method for achieving this is left open. Directives are normally implemented by national regulations made in each member state. They must be implemented by a defined date referred to in the directive.

Finally, individual nations will have some form of national (and possibly more localised) legal system in place. This may be influenced by the international legal systems covered earlier, but individual countries will have powers to implement other environmental laws as well, providing they do not contradict these two influences on the nation's legal system.

Such legal systems can differ substantially around the world. It is important that a good understanding of local legislation is gained in order to effectively manage environmental issues. The breach of environmental law is often a criminal offence and may result in a fine or a prison sentence.

As well as the criminal-law consequences, there is also the matter of compensation for those affected by environmental issues. Depending on the region/country concerned, this might involve taking legal action against the person who has caused the environmental problem through the civil legal system, and having to prove a negligent act has been carried out and was to blame for the incident.

The role of an environmentally responsible business is to ensure that, as a minimum, it complies with all the relevant environmental legal requirements. However, a progressive organisation will go beyond this and look at further improvements, all of which will help ensure its long-term success.

United Kingdom Withdrawal from the European Union

The United Kingdom withdrew from the European Union on 31 January 2020 in a process that is commonly referred to as Brexit. Key legislation covering withdrawal includes the **European Union (Withdrawal) Act 2018** and the **European Union (Withdrawal Agreement) Act 2020**.

EU law has been retained, with amendment so that it operates effectively now the UK is no longer part of the EU:

- UK legislation that implements EU requirements that was made before the end of the transition period (January 2021) has been retained.
- Certain types of EU legislation (such as EU Regulations) which are directly applicable to the UK have effect with no need for legislation to implement them.
- Case law produced by the European Court of Justice (ECJ) delivered prior to the end of the transition period remains binding on UK courts.

Following exit from the EU, the UK may choose to follow EU requirements or diverge from EU requirements.

Business Value of Environmental Achievements and Voluntary Standards

Achieving voluntary environmental standards, such as ISO 14001, adds significant value to an organisation. In some industrial sectors it is difficult to successfully tender for work without such accreditations, giving competitors who are certified a significant advantage.

> **STUDY QUESTIONS**
> 3. What are the three main reasons why organisations need to manage environmental impacts?
> 4. Identify three legal and economic effects that could occur following a pollution incident.
>
> (Suggested Answers are at the end.)

1.3 Supporting Sustainable Development

Supporting Sustainable Development

IN THIS SECTION...

- Sustainability is a concept developed at the Rio Earth Summit in 1992.
- It is about the integration of several things:
    - Environmental protection.
    - Natural resources.
    - Social progress.
    - Sustainable purchasing.
    - Competition.
    - Economic growth.
- Indicators have been established to monitor the progress of environmental objectives.

Definition of Sustainability

The natural environment provides water to drink, air to breathe, food to eat and raw materials for making things - everything we need to live healthy and productive lives.

As we strive to become wealthier, environmental protection needs to be an integral part of social and economic development. If we damage the environment, or deplete stocks of natural resources in the process, future generations will find it harder to maintain the same standard of life that we currently enjoy.

Although we all want to be healthy and prosperous, it's important to achieve this in a sustainable way that does not irretrievably damage the natural resources of the Earth. In 1992, the UN held a landmark summit meeting at which the concept of sustainable development was first widely discussed. Following the summit, the UN Rio Declaration on the Environment and Development was issued. A frequently quoted definition of sustainable development is enshrined in Principle 3 of the Declaration:

The natural environment provides everything we need to live healthy and productive lives

> "The right to development must be fulfilled so as to equitably meet developmental and environmental needs of present and future generations."

Another well known definition of sustainable development quoted in *Our Common Future*, often known as the Brundtland Report, is:

> "Sustainable development is development that meets the needs of the present without compromising the ability of future generations to meet their own needs."

Importance of Sustainable Development

Sustainable development is a phrase that is often used, but the underlying concept is not always understood. Sustainability can help an organisation develop and also contribute to global progress because it does so without depleting natural resources.

The idea of sustainable development arose from the realisation that our way of life is inherently 'unsustainable'. Environmental issues brought the first warning; we saw increasing loss of forests and wilderness areas, more animals and plants facing extinction, and the stock of fish in our oceans declining. We've also seen famines devastate huge areas in Africa, fuel shortages followed by a steep rise in fuel prices, and the increasing threat to the world's climate from global warming - all of which reinforce the idea that we are taking too much from the Earth.

We know that this situation cannot continue indefinitely. If we don't make significant changes - and soon - we will face a shortage of food, fuel, land, wild places, wild animals and plants, and other resources.

Sustainable development is about taking action to realign our economies with the capacity of the planet. There are several steps that organisations can take towards sustainability; they can endeavour to:

- Protect the environment.
- Use natural resources wisely.
- Strive for social progress.
- Use sustainable purchasing practices.
- Be competitive.
- Maintain stable economic growth.

Protecting the Environment

Our natural environment provides us with water, air, land to grow food, mineral resources, space and pleasant surroundings in which to live. Organisations that take effective steps to identify the potential environmental impacts of their activities and to protect the environment are therefore making an important contribution to sustainability. Key actions might include:

- Implementing an environmental management system.
- Designing products and processes that create less pollution.
- Meeting, or exceeding, the requirements of environmental pollution legislation.

Using Natural Resources Wisely

We've known for a long time that if we manage the naturally **renewable resources** available to us - such as fish in the sea, or timber in forests - we can maintain a sustainable supply. We can regularly harvest a proportion of the resource without depleting the supply - because that proportion will be replaced by natural growth. However, through modern fishing methods and indiscriminate clearing of woodland and forests for the growth of crops, we have gone far beyond the proportion of these resources which can be naturally restored. As a result, the supply of some natural resources is becoming very limited.

We also depend on many **non-renewable resources**, and using these in a sustainable way can be even more challenging. For example, one non-renewable resource we depend on heavily is oil. We use oil for fuel, and to make other materials, such as plastics.

There are several steps organisations can take to manage their use of non-renewable resources more sustainably:

- Using less energy.
- Designing products and processes that use fewer non-renewable-based (e.g. oil-based) resources.
- Finding renewable alternatives to the use of non-renewable resources (e.g. oil).
- Recycling non-renewable-based (e.g. oil-based) materials, such as plastics, wherever possible.

Achieving Social Progress

In addition to the environmental and economic needs of mankind (present and future), sustainable development should also consider social well-being.

It's been proposed that a range of factors should be taken into account when measuring sustainability in the UK, including:

- Life expectancy.
- Poverty.
- Long-term unemployment.
- Social mobility.
- Housing provision.

1.3 Supporting Sustainable Development

We live and operate in a global market, so it's important to consider this concept on a wider scale. It's unreasonable to expect to achieve prosperity and high levels of well-being in the developed world if we cannot do so without decreasing the prosperity and well-being of workers in less developed areas. International trade must be fair and ethical, promoting social goals in the developing world.

There are steps organisations can take to ensure they're not contributing to poor conditions for workers, including ensuring:

- They do not purchase goods that are produced using child labour.
- Their workers all work in safe, healthy environments and receive fair wages.

Being Competitive

There are various ethical arguments in support of adopting a sustainable development path. However, it's not just an ethical issue; being responsible and accountable can also have significant commercial advantages. Awareness of, and concerns related to, environmental protection and ethical trade are becoming widespread. It's important for organisations to react to this and recognise that:

- There's a huge investment market which will only consider ethically sound companies. These Socially Responsible Investment (SRI) funds use corporate responsibility performance measures to help them choose well-run companies to invest in.
- Customers not only understand, but also care about, environmental and fair-trade issues; they're willing to support initiatives such as the Forest Stewardship Council (FSC), which promotes products made from wood grown in sustainable forests.

As a result, organisations can benefit from a sustainability strategy for a number of reasons. They may enjoy:

- Reduced operating costs, e.g. as a result of increased energy efficiency and reduced waste production and water usage.
- Better access to capital, e.g. from investors who value sustainable business practices and recognise the advantages of these policies.
- Improved reputation, and therefore sales, from customers who appreciate the importance of supporting sustainable business and want to be confident that the products they purchase have been produced in an environmentally and socially responsible way.

Maintaining Stable Economic Growth

The economic argument tends to be very important to organisations, but economic growth is also important to our well-being. It's important to us to be able to enjoy life, in addition to meeting our more basic needs for clean water and food, etc. Ensuring stable economic growth for the future is essential for our own well-being and the well-being of future generations, but this can raise serious questions regarding sustainability.

Some believe that growth implies that we will continue to consume more and more resources - which is incompatible with the central objective of sustainability. Others believe this is a misunderstanding and 'economic growth' should be re-defined to avoid the implications associated with the current definition, which is an increase in monetary Gross Domestic Product (GDP).

It's been suggested that a better target in terms of a sustainable economy would be to change our lifestyle, by:

- Consuming fewer resources.
- Having less money, and relying on stronger communities and less stressful lifestyles.
- Reducing the financial inequality across all members of our society.

Corporate Social Responsibility

Corporate Social Responsibility (CSR) is an organisational approach that is very closely aligned with the concepts of sustainability. Organisations that pursue CSR seek to embed social, environmental and ethical management at the heart of their businesses. CSR requires that an organisation should be accountable to its stakeholders - customers, investors, employees, suppliers, local communities and society as a whole - for managing its social, environmental and wider economic impacts.

Many companies now produce regular CSR reports that cover the three main strands of environmental, social and economic sustainability:

Environmental	Resource consumption
	Control of pollution
	Energy and climate change
	Biodiversity
	Supply chain impacts
Social	Working conditions
	Fair wages
	Diversity
Economic	Socially responsible investment
	Fair contracts and pricing
	Trading with emerging economies
	Taxes and subsidies

A consensus is emerging on good reporting practice, and standards have now been developed to guide reporting organisations, notably the international Global Reporting Initiative.

> **MORE...**
>
> Recent examples of corporate environmental reports, together with the latest reporting guidelines, can be found on the Global Reporting Initiative (GRI) website:
>
> https://www.globalreporting.org/

The Business Case for Sustainable Development

Specifically, considering sustainability and incorporating sustainability considerations into core business decisions can benefit organisations through:

- Resource efficiency - evidence now points to the fact that reduction of resources needed to run a business will lead to higher business growth than 'business as usual', while at the same time reducing pressure on the environment and enhancing employment.

- Impacts of climate change - climate change is set to impact on all our lives by creating unpredictable weather patterns, leading to a shift in the availability of certain raw materials.

- Risk management - adapting to changing global conditions and creating opportunities and value within planetary limits. In our resource- and carbon-constrained world, a new framework for business decision-making is evolving where ecological limits are paramount and will be key success criteria for future business operations.

- Attracting and retaining quality employees - facilitated by maintaining an ongoing sustainability improvement programme, identifying employees as major stakeholders.

- Managing and enhancing reputation - this has been a major driver for many corporations engaging in Corporate Social Responsibility (CSR), and the same principles apply to any size organisation.

- Stakeholder engagement - improving relations with key partners. Identifying appropriate communication methods to engage with priority stakeholders will help the ongoing success of an organisation by ensuring relationships are optimised to meet the exact requirements of the organisation and stakeholder, from suppliers to customers.

1.3 Supporting Sustainable Development

Role of the United Nations' Sustainable Development Goals (SDGs)

Rio+20 Conference on Sustainable Development

The Rio+20 Earth Summit in June 2012 signed off a plan that will help put sustainable development on the international agenda. Key points include:

- Commitment to develop Sustainable Development Goals (SDGs) covering the elements of sustainable development (social, economic and environmental).
- Recognition of the key role of the green economy as a massive economic opportunity that countries should adapt to in the future.

2030 Agenda for Sustainable Development

A set of SDGs and targets was developed at the United Nations Sustainable Development Summit in September 2015. They were developed for the general categories people, planet, prosperity, peace and partnership. The goals must be achieved by 2030 and are:

"Goal 1. End poverty in all its forms everywhere

Goal 2. End hunger, achieve food security and improved nutrition and promote sustainable agriculture

Goal 3. Ensure healthy lives and promote well-being for all at all ages

Goal 4. Ensure inclusive and equitable quality education and promote lifelong learning opportunities for all

Goal 5. Achieve gender equality and empower all women and girls

Goal 6. Ensure availability and sustainable management of water and sanitation for all

Goal 7. Ensure access to affordable, reliable, sustainable and modern energy for all

Goal 8. Promote sustained, inclusive and sustainable economic growth, full and productive employment and decent work for all

Goal 9. Build resilient infrastructure, promote inclusive and sustainable industrialisation and foster innovation

Goal 10. Reduce inequality within and among countries

Goal 11. Make cities and human settlements inclusive, safe, resilient and sustainable

Goal 12. Ensure sustainable consumption and production patterns

Goal 13. Take urgent action to combat climate change and its impacts*

Goal 14. Conserve and sustainably use the oceans, seas and marine resources for sustainable development

Goal 15. Protect, restore and promote sustainable use of terrestrial ecosystems, sustainably manage forests, combat desertification, and halt and reverse land degradation and halt biodiversity loss

Goal 16. Promote peaceful and inclusive societies for sustainable development, provide access to justice for all and build effective, accountable and inclusive institutions at all levels

Goal 17. Strengthen the means of implementation and revitalise the Global Partnership for Sustainable Development

* Acknowledging that the United Nations Framework Convention on Climate Change is the primary international, intergovernmental forum for negotiating the global response to climate change."

Source: *Transforming our world: the 2030 Agenda for Sustainable Development*, United Nations, 2015

STUDY QUESTION

5. Explain why achieving sustainable development is difficult.

(Suggested Answer is at the end.)

The Role of National Governments and International Bodies in Formulating a Framework for the Regulation of Environmental Management

IN THIS SECTION...

- There is an international framework for environmental law, as some environmental issues require international co-operation.
- 'Best available technique' is a key principle in pollution assessment and control.
- Enforcement agencies play a key role in environmental management and often have powers to issue permits, inspect sites, serve notices and prosecute.

International Law

International environmental law comprises the body of rules derived both from international agreements and customary international law to which sovereign states have expressly or implicitly (via state practice) consented.

International requirements are not directly law in a country that has agreed to follow them. They must be made into national law and as such are a major influence on national environmental laws and standards.

International environmental law is often in the form of **Conventions**, which are legally binding agreements between a significant number of different states, often developed under the auspices of the United Nations. Conventions are often implemented through subsidiary **Protocols**. The **Vienna Convention for the Protection of the Ozone Layer (1985)**, for example, is implemented through the various amendments to the **Montreal Protocol on Substances that Deplete the Ozone Layer,** which sets the detailed timetable for phasing out production of ozone-depleting substances. A number of the most significant international Conventions and their subsidiary Protocols are summarised in the table that follows.

Climate change

The Role of National Governments and International Bodies in Formulating a Legal Framework for the Regulation of Environmental Management 1.4

International conventions and protocols on the environment

Subject	Convention	Subsidiary Protocols
Climate change	United Nations Framework Convention on Climate Change 1992	Kyoto Protocol 1997 Doha Amendment 2012
Protection of the ozone layer	Vienna Convention for the Protection of the Ozone Layer 1985	Montreal Protocol 1989 Latest revision - Beijing 1999
Air pollution	Convention on Long-Range Transboundary Air Pollution ('Geneva Convention') 1979	Oslo Protocol 1994 Aarhus Protocol 1998 Gothenburg Protocol 1999
Marine pollution	Convention for the Protection of the Marine Environment of the North-East Atlantic ('OSPAR Convention') 1992	
Hazardous wastes	Convention on the Control of Transboundary Movements of Hazardous Wastes and Their Disposal ('Basel Convention') 1992	Basel Protocol 1999
Persistent organic pollutants	Convention on Persistent Organic Pollutants ('Stockholm Convention') 2001	
Habitat protection	Convention on Wetlands of International Importance ('Ramsar Convention') 1971	Paris Protocol 1982

We will now take a look in more detail at a number of these key international legal requirements.

Climate Change

> **DEFINITION**
>
> **KYOTO PROTOCOL**
>
> Is an international agreement looking at reducing greenhouse gas emissions. Developed countries must reduce greenhouse gas emissions, whereas developing countries are only required to monitor and report emissions.

The United Nations Framework Convention on Climate Change (UNFCCC) is a treaty that entered into force in 1994 and is the foundation stone of international efforts to tackle climate change. There are now 195 parties to the Convention, including the European Union and the UK. The Convention is implemented by the **Kyoto Protocol**.

The main objective of the Protocol is to reduce emissions of six greenhouse gases: carbon dioxide, methane, nitrous oxide, sulphur hexafluoride, hydrofluorocarbons and perfluorocarbons. The first phase ('first commitment period') of the implementation of the Protocol ran from 2008 to 2012 and set national emission targets, which, overall, amounted to an average 5% emissions reduction of this basket of gases compared with 1990 levels. Developing countries, such as Brazil, had no emissions reduction targets but committed to monitor and report their emissions.

1.4 The Role of National Governments and International Bodies in Formulating a Legal Framework for the Regulation of Environmental Management

The **Doha Amendment** to the Kyoto Protocol was adopted in 2012 and introduced a second commitment period running from 2013 to 2020. Participants agreed to reduce greenhouse gas (GHG) emissions by at least 18% against 1990 levels in this period. Countries that signed up to the second commitment period (so-called 'Annex l' countries) have flexibility in the actions they take to achieve the specified reductions.

The adoption of the Kyoto Protocol was highly controversial, with certain major trading countries resisting the concept of binding emissions-reduction targets. The Annex I countries accounted for only about 15% of global GHG emissions, and a number of important industrialised nations, including the USA, Canada and Russia, did not sign up to the second commitment period of the Protocol.

The **Paris Agreement** was developed at the Paris climate conference in December 2015 and formally entered into force in November 2016 when a sufficient number of countries (representing at least 55% of the world's GHG emissions) had ratified the agreement. The agreement entered into force in 2020.

Key aspects of the agreement:

- Participant countries will have to review climate plans on a regular basis and ensure that action is taken to deal with climate change.
- Development of an aim of net zero emissions by the end of the century.
- A specific legal requirement to reduce emissions on a five-yearly basis from 2025. All participant countries must be independently reviewed for progress towards their emission reduction pledges.
- Developed and emerging economies must mobilise $100 billion per year from public and private funding to assist vulnerable and poor countries in protecting themselves against the consequences of climate change.

The Conference of the Parties to the UN Framework Convention on Climate Change (UNFCCC) (COP24) was held in December 2018 in Katowice, Poland. The key outcome of the conference was to agree and adopt a series of decisions to ensure the implementation of the Paris Agreement. This is known formally as the Paris Agreement Work Programme (PAWP), or more informally as the Paris rulebook.

The Conference of the Parties to the UN Framework Convention on Climate Change (UNFCCC) (COP26) was held in October/November 2021 in Glasgow, UK. Key outcomes of the conference include:

- Timetable to produce more ambitious National Determined Contributions (NDCs). Current NDCs are considered inadequate to meet the limit of temperature rise to 1.5 degrees C.
- Commitment to phase down the use of coal for power generation.
- Nations are 'invited' to reduce methane emissions this decade.
- Phase-out of subsidies that artificially lower the price of fossil fuels.
- Glasgow climate pact agrees to double the proportion of climate finance to assist developing countries to adapt to climate change.
- Developed countries are urged to fully deliver the originally agreed $100bn goal of funding to developing countries by 2025.
- Carbon offsetting that is used to reach NDCs should rely on 'real, verified and additional' emissions removal from 2021.

The **Kigali Amendment** to the **Montreal Protocol** sets legally binding targets that specify the phased reductions of the hydrofluorocarbons (HFCs, a potent group of fluorinated greenhouse gases). HFC reductions are measured on overall CO_{2e} (carbon dioxide equivalent). The Kigali Amendment entered into force on 1 January 2019.

The Role of National Governments and International Bodies in Formulating a Legal Framework for the Regulation of Environmental Management | 1.4

The HFC phase-down schedule under the Kigali Amendment is provided in the table below.

HFC Phase-Down Schedule under the Kigali Amendment

	Non-A5 (developed countries)	A5 (developing countries) Group 1	A5 (developing countries) Group 2
Baseline HFC component	2011-2013 (average consumption)	2020-2022 (average consumption)	2024-2026 (average consumption)
Baseline HCFC component	15% of baseline	65% of baseline	65% of baseline
Freeze	-	2024	2028
1st step	2019 - 10%	2029 - 10%	2032 - 10%
2nd step	2024 - 40%	2035 - 30%	2037 - 20%
3rd step	2029 - 70%	2040 - 50%	2042 - 30%
4th step	2034 - 80%	-	-
Plateau	2036 - 85%	2045 - 80%	2047 - 85%
Notes	Belarus, Russian Federation, Kazakhstan, Tajikistan, Uzbekistan, 25% HCFC component and 1st two steps are later: 5% in 2020, 35% in 2025	Article 5 countries not part of Group 2	GCC (Saudi Arabia, Kuwait, United Arab Emirates, Qatar, Bahrain, Oman), India, Iran, Iraq, Pakistan

Source: Adapted from eia-international.org/wp-content/uploads/EIA-Kigali-Amendment-to-the-Montreal-Protocol-FINAL.pdf

Ozone Depletion

The **Vienna Convention 1985** required nations to take appropriate measures to protect people and the environment against the impacts resulting from human activities that modify, or are likely to modify, the ozone layer. The main aim of the Convention is to ensure that countries undertake research, exchange information and monitor CFC production. The Convention acts as a framework for international efforts to combat ozone depletion.

More specific requirements on the banning and phasing out of Ozone-Depleting Substances (ODSs) are present in the **Montreal Protocol on Substances that Deplete the Ozone Layer 1987**. Since it initially opened for signatures in 1987 (coming into effect in 1989), the Protocol has undergone seven revisions. This has resulted in the phase-out dates for ODSs as identified in the table that follows.

Ozone-Depleting Substance	Developed Countries Phase-Out Dates	Developing Countries Phase-Out Dates
CFCs	1995	2010
Halons	1993	2010
Methyl Chloroform	1995	2015
Carbon Tetrachloride	1995	2010
HCFCs	2020	2030
HBFCs	1995	1995

1.4 The Role of National Governments and International Bodies in Formulating a Legal Framework for the Regulation of Environmental Management

> **MORE...**
>
> Further information on the Montreal Protocol can be found at:
>
> https://ozone.unep.org/treaties/montreal-protocol

Basel Convention on the Control of Transboundary Movements of Hazardous Waste and their Disposal

As we saw earlier, this law aims to prevent or control the transfer of hazardous waste and certain types of plastic wastes from developed countries to less developed countries. It aims to control international waste movements by requiring that exporters of waste receive written consent from importing nations prior to the waste being transported. The Convention also places a ban on the exportation or importation of waste between those who are party to the Convention and those who are not. There are, however, exceptions where a country is subject to a treaty or similar that does not undermine the Basel Convention. For example, the USA is a non-party to the Convention but has developed a number of agreements allowing for shipping of hazardous waste to those countries that have ratified the Basel Convention.

> **MORE...**
>
> As stated earlier, more information on the Basel Convention is available at:
>
> www.basel.int

OSPAR Convention 1992

The **Convention for the Protection of the Marine Environment of the North-East Atlantic** (the **OSPAR Convention**) entered into force in 1998 and has been ratified by 15 countries and the European Union (under **Decision 98/249/EC**). It works by identifying threats to the marine environment in the north-east Atlantic and programmes and measures that ensure national action is taken to combat them. It sets internationally agreed goals and monitors them to ensure that countries are complying with them.

This part of the Atlantic is under intense pressure with regard to the marine ecosystem, pollution, maritime activities, oil and gas extraction and nuclear energy. Achievements since the Convention's introduction include:

- Significant reduction in phosphorus and heavy metals.
- Reductions in discharges from nuclear plants.
- Ban on waste dumping.
- Network of OSPAR marine-protected areas.

> **MORE...**
>
> More information on the OSPAR Convention is available at:
>
> www.ospar.org

The Role of National Governments and International Bodies in Formulating a Legal Framework for the Regulation of Environmental Management

The Importance of Knowing and Understanding Local Legislation

Environmental legislation can be significantly different in countries around the world. It may even differ within a country. It is therefore important that a strong knowledge of local legislation is developed to ensure compliance with relevant environmental law.

It is important therefore for multinational organisations to ensure that environmental laws are understood and complied with in the locations where they operate as requirements could be different.

Meaning of BAT and BPEO

Best Available Techniques (BAT) applies to industrial installations that require an integrated pollution prevention and control permit.

In the context of BAT:

- **'Best'** means, in relation to techniques, the most effective in achieving a high general level of protection of the environment as a whole.

- **'Available Techniques'** means those techniques that have been developed, on a scale which allows implementation in the relevant industrial sector, under economically and technically viable conditions, taking into consideration the cost and advantages, whether or not the techniques are used or produced inside a country (as long as they are reasonably accessible to the operator).

- **'Techniques'** includes both the technology used and the way in which the installation is designed, built, maintained, operated and decommissioned.

A related concept is that of **Best Practicable Environmental Option (BPEO)**, which has very wide application. It is a holistic approach like BAT, aimed at avoiding situations in which narrow thinking to limit pollution to one medium could, in fact, increase pollution to another medium. Therefore, the environment as a whole is considered. The concept can be defined as:

> *"the outcome of a systematic consultative and decision-making procedure which emphasises the protection and conservation of the environment across land, air and water. The BPEO procedure establishes for a given set of objectives, the option that provides the most benefits or the least damage to the environment, as a whole, at acceptable cost, in the long term as well as the short term."*

(Source: UK Royal Commission on Environmental Pollution)

The Role of Enforcement Agencies and Consequences of Non-Compliance

There is no harmonised global standard for the enforcement of environmental law, so legal and enforcement systems vary between countries. There are, however, some general principles that normally apply.

Each country or region has one (or more) enforcement agencies responsible for enforcing environmental law. Such an agency is effectively the 'environmental police force'. In some circumstances, the agency may be, or may enlist the help of, the national or regional police.

These agencies often:

- Provide advice.
- Investigate environmental incidents.
- Take formal enforcement action to force organisations to comply with the law.
- Start criminal proceedings against persons or organisations they believe have committed offences.

1.4 The Role of National Governments and International Bodies in Formulating a Legal Framework for the Regulation of Environmental Management

Enforcement agencies are also often involved in authorising activities that can have an impact on the environment. This usually takes the form of an environmental permit (depending on the legal system of a country, this may alternatively be called a licence or authorisation, as there is no worldwide standard terminology used). An environmental permit will set conditions on activities that could impact on the environment. For example, in many countries before an organisation may discharge potentially polluting materials to rivers, streams and other types of watercourses, they will have to apply for a permit. If the permit is granted it will set limits on polluting parameters associated with the discharge, such as pH, suspended solids and heavy metal, etc. Other activities that are often permitted in many countries include:

- Discharge to groundwater.
- Keeping, treating and disposing of waste.
- Emissions of pollutants to air.
- Integrated permits (where more than one activity is controlled by a single permit).

An enforcement agency will often be tasked with undertaking inspection of workplaces to check compliance with environmental legislation. However, as resources are often limited for governments to fund such inspections, it is often the case that sampling is used. In this context, organisations that present a high environmental risk owing to the activities that they undertake will be inspected more frequently than those that present a low environmental risk.

Consequences of Non-Compliance

A breach of environmental legislation is usually a criminal offence - wherever you are in the world. Failure to meet legal standards might lead to:

- Formal enforcement action: an enforcement agency might force an organisation either to make an improvement within the workplace or to stop carrying out high-risk activities through the issue of a notice. The notice will usually state what action is needed and the timeframe for implementation. Failure to comply with a formal enforcement notice is usually considered to be an offence in itself.
- Prosecution of the organisation in the criminal courts: successful prosecution might result in punishment in the form of a fine.
- Prosecution of individuals, such as directors, managers and workers: successful prosecution might result in punishment in the form of a fine and/or imprisonment.

As stated earlier, as well as the criminal-law consequences, there is also the matter of compensation for those affected by an environmental incident. Depending on the region/country concerned, this might involve:

- Those wronged taking legal action through the civil legal system and having to prove that the organisation had been negligent and is to blame for the incident.
- Claiming compensation from national or regional compensation schemes, with no requirement to prove negligence or blame through the use of the legal system.

STUDY QUESTIONS

6. State the international laws that cover:
   (a) Protection of the ozone layer.
   (b) Transfer of hazardous waste.
7. Define 'BPEO'.
8. Around the world, what types of activities are environmental permits often required for?

(Suggested Answers are at the end.)

Summary

This element has dealt with some of the basic principles of environmental management.

In particular, this element has:

- Explained that environmental management covers a wide range of disciplines, e.g. physics, chemistry, biology, geology and engineering but also town and country planning, public health and legislation.
- Outlined organisational barriers to good standards of environmental management, including the complexity of the environment, conflicting and competing demands, and behavioural issues.
- Described key environmental issues of concern, including local effects of pollution; carbon emissions and climate change; air pollution and the ozone layer; water resources; deforestation; soil erosion and land quality; land grabbing; energy supplies; waste disposal and the impacts of agriculture.
- Explained three main reasons for an organisation to actively and positively manage its environmental impacts:
    - Moral/ethical - 'doing what is right'.
    - Legal - importance of complying with relevant legislation.
    - Financial - the 'polluter pays' principle provides a range of financial tools to encourage organisations to account for pollution through the balance sheet.
- Referred to the pillars of sustainable development: environmental protection, natural resources, social progress, sustainable purchasing, competition, economic growth.
- Explained how enforcement bodies work with industry to prevent and reduce pollution. Through a network of inspectors, they are responsible for inspecting processes and sites that are permitted, in order to ensure permit conditions are complied with.
- Outlined the options available to the regulating authorities when enforcing legislation, including use of notices and prosecutions.

Exam Skills

Introduction

To pass the NEBOSH Environmental Management Certificate, you need to perform well during your 'open book' exam. An 'open book' exam means exactly what the name suggests: you can access your study text and other materials during the exam. However, it will require more than simply copying text to answer the exam questions; you will need to demonstrate that you understand the subject matter and can apply the topics appropriately – this needs preparing for.

Here we will consider some practical guidelines that can be used to increase success in your 'open book' exam. The NEBOSH Environmental Management Certificate 'open book' exam differs from the traditional question-and-answer-type format and will instead consist of a realistic workplace scenario which may describe a developing situation such as an incident or environmental intervention, and you may be asked to assume a particular role, for example an environmental manager. You will then be asked to carry out some tasks. Each task will consist of one or more questions, where your answers will be relevant to the scenario.

Wider Reading, Research and References

During the 'open-book' exam you are allowed to access many different types of resources such as RRC study text and revision guides, professional journals, digital resources on the Internet such as Environment Agency guidance, bilingual dictionaries, etc. Remember though that this is not a substitute for thorough study of your course materials and specific preparation in the form of revision to ensure that you are familiar with all the topic areas included in the syllabus. NEBOSH "will not be looking for anything from your answers that has not been covered in the Unit syllabus".

NEBOSH has designed the 'open-book' exam "to assess the same learning outcomes to the same level" as the previous method of examination. It will "measure depth of understanding" and require you to apply your knowledge to the scenario given rather than simply recalling memorised information. It is therefore "very important that you prepare and revise" for the assessment.

Wider Reading and Research

As part of your preparation, NEBOSH recommends that you should:

- "Complete wider reading so you don't have to do this during the examination."
- Conduct "wider research to understand how your studies relate to the real world".

To help with this you should make use of the **MORE...** boxes included in your course material. These highlight additional resources related to the topics you've been studying. Reading them will assist you in increasing the depth of your understanding and applying it to real life situations. The links in **MORE...** boxes are current at the time of publication of your materials but may have changed since, e.g. due to an item being moved elsewhere, being replaced or perhaps withdrawn from publication altogether.

> **HINTS AND TIPS**
>
> If the link in a **MORE...** box no longer works, you can search for the item in question by entering its title in a search engine such as Google or Bing. This should lead you to its current location if it's still available on the Web. In the event that it's no longer available, your tutor may be able to suggest a replacement.

Your wider reading may also include research using other resources mentioned in your course materials or resources that you've found yourself, perhaps by entering a topic name in a search engine. Always be careful though to use only authoritative sources for your research, such as Government or official websites, and to safeguard the security of your personal information.

Exam Skills

Making Use of Your Resources

NEBOSH states that "You are expected to offer your own analysis and presentation of information gained from your research". So you need to:

- Organise your notes in advance so that you can find things quickly.
- Use your own words to express what you've learned and apply the knowledge you've gained.

You are permitted though to make reference to sources, ideas or work of other people or organisations. Where you do so, it's important to acknowledge where that information comes from because your open-book exam response must be your own work and you don't want to be guilty of plagiarism and subject to investigation by NEBOSH.

> **DEFINITION**
>
> **PLAGIARISE**
>
> "To steal ideas or writings from another person and present them as one's own".
>
> (Source: The Chambers Dictionary)

References

Direct Citations

You may want to quote specifically from a publication you've accessed. Here you should put the quote in inverted commas and make it clear where it's from. (See also **Reference List** below.)

Example:

ISO 14001 defines an environmental impact as: "change to the environment, whether adverse or beneficial, wholly or partially resulting from an organisation's environmental aspects".

Indirect Citations

You may just want to refer to information that you've read without quoting from it directly. In this case, express it in your own words but make sure that you identify the source. (See also **Reference List** below.)

Example:

ISO 14001 defines what is meant by an environmental impact.

Reference List

Your exam answers should include a list of the materials you've referred to – those you've quoted from, those you've mentioned and any other materials that you looked at during the exam. This doesn't count towards the word limit for your answers and it would be best to make the list as you write your answers to ensure that the list is complete.

We don't recommend using the Harvard (or any other) referencing system for this purpose because NEBOSH has made it clear that a simple list is all that is required.

Your list should be at the end of your answer document and should include:

- the title of the publication/document,
- the author,
- the year of issue, and
- a Web address (for Internet publications.

Example:

ISO 14001:2015 Environmental management systems - Requirements with guidance for use, International Organisation for Standardisation, 2015

Exam Skills

Guidance on applying the Waste Hierarchy, DEFRA, 2011 (https://assets.publishing.service.gov.uk/government/uploads/system/uploads/attachment_data/file/69403/pb13530-waste-hierarchy-guidance.pdf)

(In the above Example:

- ISO 14001:2015 is included because quoted from and referred to in the text.
- Guidance on applying the Waste Hierarchy is included because consulted during the exam.)

Note: If you use the NEBOSH answer sheet, there is a place there for your references.

Exam Technique

The 'open-book' exam will test you on your ability to "demonstrate analytical, evaluation and creative skills as well as critical thinking" and how you apply your learning to your answers. In other words, you will need to show what you can do with your knowledge to solve the problems presented to you – and this may take practice. To assist you in showing your knowledge, let us look at a step-by-step approach that you can adopt when answering your exam questions.

Step 1 Read the scenario - the first step is to read the scenario carefully. Take care with this as it is very easy to misread words in the rush to get writing. It's likely you will read the scenario more than once.

Step 2 Look at the first task - a task is an activity or piece of work that will be part of a larger project. The task may be split up into several sub-tasks. The task is used to indicate the questions. Read each question carefully. This is what your knowledge is being tested on - so your answer must address the question. Demonstrating knowledge alone will not gain marks - your knowledge has to be applied to the task - so you need to clearly understand the question being asked.

Step 3 Look at the marks - each task or sub-task, question or part of a question will have the maximum number of marks indicated in brackets. For each mark to be awarded, the examiner will expect a piece of information that demonstrates an analytical evaluation of the task that has been set. The marks available give an indication of how much you will need to write and, to a lesser extent, how long you should spend on this part of each question.

Step 4 Re-read the scenario and task question - to check that you have properly interpreted them and understood them. There are no marks available for answering the task question that you think you see rather than the one that the examiner asked you.

Step 5 Draw up a plan - this can take the form of a list or a mind map that helps you unload information quickly and make sure you have enough factors (or things) in your answer to gain the available marks. Jotting down a plan can help you remember key points. The plan is also your aide-mémoire to keep you on track as you start to write your full answer. Your plan can use information from the scenario to support your answer. When it comes to the exam, there is no need to submit your plan.

NEBOSH has said that "You can expect to see questions that ask What? Could? How? Why? and Where? " but "Whatever format the question, it will relate to the syllabus and learning outcomes of your qualification".

When writing your answer, you must ensure that the structure of the task appears in the structure of your answer. So, for example, if the task question has a part (a) and a part (b), your answer must follow the same structure. Answer part (a) and label it clearly for the examiner as the answer to part (a). Then leave a gap (one line will do) and answer part (b) and label it clearly. The examiner must be able to see the two separate parts of your answer and it must be clear to them which parts are the answer to which questions. One long paragraph of text that contains all parts of the answer jumbled up together cannot gain full marks, even if all of the relevant information is there.

The above exam technique is tried and tested and is the best way to approach each exam task. 'Open-book' exams are a new approach for NEBOSH but the education sector has used 'open-book' exams for a long time.

NEBOSH gives a 24-hour window of time for the exam to reflect different time zones learners work and live with. You are not expected to spend 24 hours completing the exam. You will, however, need to monitor your progress on completing all the tasks and you should plan for drafting answers, reflection and amending answers, building in time for breaks.

Remember, too, that there will be an overall word count for the exam. It's important not to exceed the recommended total by more than 10% so do allow time to check this

Exam Skills

Practice Exam Questions

At the end of Elements 1 to 9 there is a practice exam question for you to attempt, with guidance on how to approach the question in addition to an example of how the question could be answered.

Remember that when answering exam questions, information from additional reading and personal experience can be included. Examining bodies encourage this and it will enhance your answers.

The study text provides some useful links to external sources - look out for the 'MORE...' boxes within the materials - these contain useful links to relevant topics.

Please feel free to contact your tutor if you have any queries or need any additional guidance.

Exam Skills

Question

Taking into account what we have just covered on exam technique, consider the following exam-style question.

> ### Scenario
> Following a customer audit, an organisation has decided to appoint you as an environmental adviser. The organisation has no formal approach to managing environmental issues. While carrying out a review, you find a significant amount of scepticism from many employees regarding effective environmental management. A meeting with yourself and the Director has been arranged to discuss the issue.
>
> ### Task: Barriers to Effective Environmental Management
> To enable you to understand a way forward and to change employees' opinions on environmental management, prepare for the meeting by completing some notes on the three key barriers that organisations often encounter when wanting to improve environmental performance. **(6 marks)**

Approaching the Question

Think about the steps you would take to answer the question:

- Read the scenario carefully. With this question you need to develop an idea of the three key barriers to improvements in environmental performance. Note: the question states 'three barriers' so it is likely that the barriers are quite specific to the NEBOSH EMC syllabus.
- Now look at the task – prepare notes on:
  - The barriers to environmental management covered in the course.
- Consider the marks available. In this case, there are 6 marks available for 3 barriers, so there are 2 marks for each barrier. You cannot just state 6 items of information on one barrier.
- Read the scenario and task again to make sure you understand them and have a clear understanding of the barriers.
- Jot down an outline plan - this might include:
  - Complexity, conflicting demands, behavioural issues.

Now have a go at the question yourself.

Example of How the Question Could be Answered

The first barrier is complexity. Organisations and associated environmental issues can be complex; deciding which impact to tackle and ways of achieving improvement require knowledge of environmental management - particularly in terms of understanding options for improvement.

Competing and conflicting demands present another barrier. Organisations must make a profit to survive and this can be seen as the main driver for an organisation, whether or not the organisation's activities cause a significant impact to the environment.

Behavioural issues also present another challenge. The way people behave at work can be very difficult to change. As such, if change is made it can take time, and good reasons must be provided to people in order for them to make changes.

References

NEBOSH Open Book Examinations: Learner Guide - Guidance for preparing for an open book examination, NEBOSH, 2021 https://www.nebosh.org.uk/documents/open-book-examination-learner-guide

The Chambers Dictionary, Chambers, 2005

ISO 14001:2015 *Environmental management systems — Requirements with guidance for use*, International Organisation for Standardisation, 2015

Guidance on applying the Waste Hierarchy, DEFRA, 2011 (https://assets.publishing.service.gov.uk/government/uploads/system/uploads/attachment_data/file/69403/pb13530-waste-hierarchy-guidance.pdf)

Element 2

Environmental Management Systems

Learning Outcomes

- Understand the requirements of, and work within, an environmental management system, whilst contributing to continual improvement.

Learning Objectives

Once you've read this element, you'll be able to:

1. Recognise the key features and appropriate content of an effective EMS (based on the requirements of **ISO 14001**).

2. Discuss the benefits and limitations of introducing a formal EMS into the workplace.

Contents

Reasons for Implementing an Environmental Management System (EMS) — 2-3
Introduction to Environmental Management Systems — 2-3

The Key Features and Appropriate Content of an Effective EMS — 2-6
Introduction to ISO 14001 — 2-6
ISO 14001 — 2-6
Initial Environmental Review — 2-7
Context of the Organisation — 2-8
Leadership — 2-9
Planning — 2-10
Support — 2-13
Operation — 2-14
Performance Evaluation — 2-15
Improvement — 2-24
Eco-Management and Audit Scheme (EMAS) — 2-25

Benefits and Limitations of Introducing a Formal EMS into the Workplace — 2-26
Benefits of Introducing a Formal EMS into an Organisation — 2-26
Limitations of Introducing a Formal EMS into an Organisation — 2-27

Summary — 2-29

Exam Skills — 2-30

Reasons for Implementing an Environmental Management System (EMS)

IN THIS SECTION...

There are a number of reasons for implementing an Environmental Management System (EMS), including:

- Demonstrating management commitment.
- Having common management principles with other systems.
- Responding to stakeholder pressure.
- Acting in a socially responsible way.

Introduction to Environmental Management Systems

Many organisations want to improve their environmental performance, but find it difficult to know how to go about this in an effective and efficient manner. We have seen that any organisation can have many different interactions with the environment. Deciding which issues should have priority, and exactly what needs to be done to address these, is not always easy.

Environmental Management Systems (EMSs) are tools that help organisations manage their environmental issues systematically and comprehensively. The EMS model set out in the international **ISO 14001** standard is now the most widely adopted. This is based on the total quality management concepts that underpin other commonly implemented management systems (notably the ISO 9001 Quality Management System standard) and where the emphasis is on continual improvement.

EMSs help organisations manage their environmental issues

Organisations that follow this EMS approach can enjoy a number of benefits. For example, they can:

- Manage their environmental impacts in the most resource-efficient way to bring about improvements.
- Make cost savings through better control of such issues as energy consumption and waste management.
- Achieve compliance with environmental legislation.
- Demonstrate their environmental commitment to customers and other interested parties, such as regulatory authorities, insurance companies, shareholders and local residents.

Demonstrating Management Commitment

Implementing an EMS is a significant undertaking and cannot succeed without commitment from senior management because:

- It can be a time-consuming task requiring potentially significant use of resources in terms of time, money and facilities.
- The person or team implementing the EMS will need to call on expertise from many people in the organisation. If the EMS is not seen as an important issue then these people may not be willing to give the time necessary to gather the correct information. For instance, a production manager will be crucial to gaining information on the production process, the environmental impacts it may have and the potential for changes to reduce those impacts. If the production manager does not give his/her time, then valuable information will not be collected.
- There must be a clear message to the rest of the organisation that this is an organisational priority.
- The benefits to the organisation of implementing the EMS must be explained and understood by all to gain support.

2.1 Reasons for Implementing an Environmental Management System (EMS)

Successful implementation of an EMS demonstrates to the public, customers, employees and other stakeholders that management are committed to environmental performance.

Demonstrating Commitment to Pollution Prevention and Control/Protection of the Environment

An organisation that possesses a formal EMS must ensure that plans, resources, procedures and other controls are implemented to ensure protection of the environment and to prevent pollution. Indeed, there is an explicit requirement for an ISO 14001 environmental policy to commit to these requirements.

Providing a Framework for Setting Objectives and Targets

An EMS will consist of a number of objectives and targets to improve environmental performance. These are usually set to control or reduce the significant organisational environmental impacts and will often help achieve continual improvement.

> **DEFINITION**
>
> **STAKEHOLDERS**
>
> Those with an interest in your company, e.g. customers, shareholders, regulatory authorities, residential neighbours, insurance companies and the supply chain.

Corporate Social Responsibility

As we have seen, organisations that pursue Corporate Social Responsibility (CSR) seek to embed social, environmental and ethical management at the heart of their businesses. Implementing a formal EMS is wholly compatible with this approach, and an EMS can be an important element of an overall CSR programme and sustainability policies and strategies.

Sharing of Common Management System Principles with Quality and Health and Safety Management

Quality, environmental and health and safety management are all based on a common platform and share the same structure. A way to effectively manage them is by following the Plan-Do-Check-Act (PDCA) cycle (see later). It is quite possible to integrate the management of the three disciplines into a single management system, which can represent significant savings to an organisation. It is, however, highly unlikely that a single person will have sufficient knowledge and experience in all these areas to effectively manage the entire system, and you should not underestimate the need for specialist knowledge in these areas.

Reasons for Implementing an Environmental Management System (EMS) 2.1

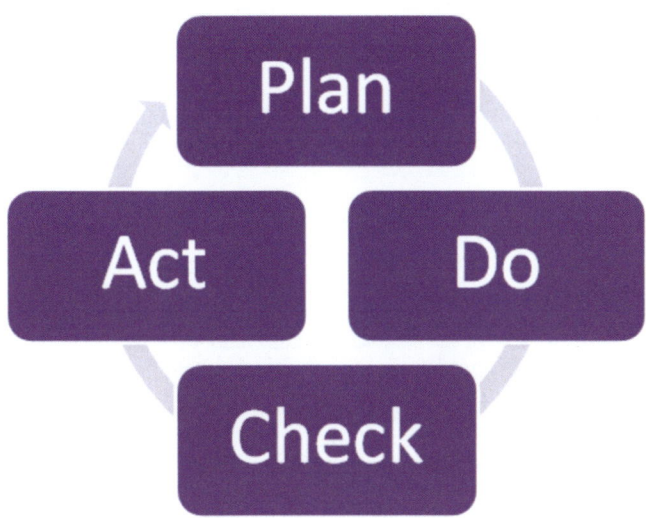

Pressure from Interested Parties

Pressure to improve environmental performance by implementing an EMS can occur from interested parties (stakeholders). These may include customers, regulators, local community and non-governmental organisations. When developing a formal EMS, the needs of such groups must be taken into account. For example, some powerful customers can exercise huge influence over suppliers - making it difficult to become an 'approved supplier' (or retain this status) unless an EMS is in place (formal or otherwise). You may be familiar with the 'approved supplier' questionnaire, where you can either painlessly tick the box on the front, confirming that you have a formal, certificated EMS, or fill in a 60-page supplier questionnaire, which asks detailed questions about the informal EMS that you may or may not have.

This 'pushing back' onto suppliers is all part of exercising greater control over the inputs into the organisation. Regulators must also be satisfied - the granting of an Environmental Permit effectively requires at least some elements of an EMS to be in place.

> **STUDY QUESTION**
>
> 1. Why is management commitment so important when implementing an environmental management system?
>
> (Suggested Answer is at the end.)

2.2 The Key Features and Appropriate Content of an Effective EMS

The Key Features and Appropriate Content of an Effective EMS

IN THIS SECTION...

- Environmental Management Systems follow standard 'Plan, Do, Check, Act' principles, with an ultimate aim of achieving continual improvement in environmental performance.
- The main components of **ISO 14001** are: context of the organisation, leadership, planning, support, operation, performance evaluation and improvement.
- Active and reactive measures for monitoring performance are available.

Introduction to ISO 14001

In this section, we're going to consider an effective environmental management system based on the requirements of **ISO 14001**.

ISO 14001:2015 is a development of earlier versions (from 1996) and aligns more closely with other standards, such as **ISO 9001** (Quality Management Systems).

In common with nearly all management systems, the **ISO 14001** standard follows the basic steps of the so-called Deming cycle (**Plan-Do-Check-Act**). This cyclical process is designed to ensure a process of continual action and improvements towards a set of objectives.

ISO 14001

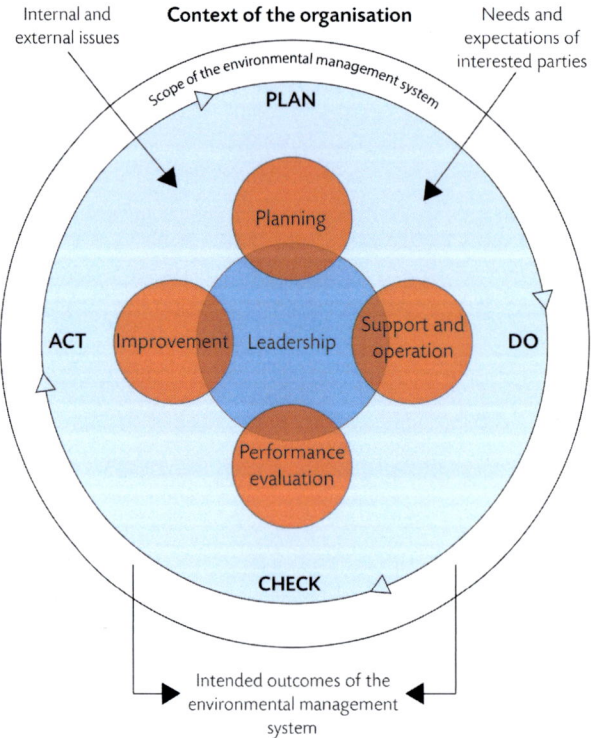

ISO 14001 management system process

The Key Features and Appropriate Content of an Effective EMS | 2.2

> **TOPIC FOCUS** ~~steps~~ features
>
> The key ~~steps~~ in the implementation of an EMS that conforms to the **ISO 14001** standard are:
>
> 1. **Context of the Organisation**
>
>   An organisation must understand all relevant issues that may affect or be affected by the organisation. The scope of the EMS must be determined and documented.
>
> 2. **Leadership**
>
>   Top management must demonstrate leadership. Roles and responsibilities should be assigned and a compliant environmental policy produced.
>
> 3. **Planning**
>
>   Key action includes the identification of relevant aspects and impacts of the organisation on the environment. The organisation must also set objectives and understand its compliance obligations.
>
> 4. **Support**
>
>   Resources must be made available to develop and operate the EMS. Requirements are also stated for competency, communication and documented information.
>
> 5. **Operation**
>
>   Consistent with a life-cycle perspective the organisation must develop operational controls. Emergency plans must also be developed and tested where appropriate.
>
> 6. **Performance Evaluation**
>
>   An organisation must monitor, measure, analyse and evaluate its environmental performance. Requirements are also present to develop and implement internal audit programmes. Top management must also review the EMS at suitable intervals.
>
> 7. **Improvement**
>
>   The cause of non-conformances must be eliminated. Non-conformances should be prevented from happening again.

Initial Environmental Review

Where there is no existing EMS, it is recommended that an initial review is undertaken to establish the organisation's current position with regard to environmental management. The review should cover all of the activities of the organisation that are within the scope of the management system, for example:

- The main and ancillary processes.
- Transport to, on and from the sites.
- Raw and intermediate material handling, storage and transfer activities.
- Waste-material storage, transfer and disposal arrangements.
- Maintenance activities.
- Packaging and warehousing.
- Procurement services and the engagement and supervision of contractors and service providers.
- The planning and design of new buildings, equipment, processes, products and services.

The review should cover four key areas:

1. The identification of environmental aspects and the evaluation of significant environmental impacts and liabilities.
2. The identification of compliance obligations (these might be legal requirements and other requirements, e.g. industry codes of practice; customer requirements; holding company requirements, etc.).
3. Examination of existing environmental management practices and procedures.
4. Assessment of previous incidents, complaints and non-conformances.

The review may also include:

- An assessment of potential commercial benefits, such as reduced energy, waste disposal and raw material costs.
- The views of key stakeholder groups (e.g. local residents, regulators, major shareholders).

The review should enable an organisation to answer the question: 'Where are we now?'

Context of the Organisation

Initially, an organisation must understand its context. This includes positive and negative internal and external relevant issues such as the environmental conditions that may be affected by the organisation or environmental conditions that could affect the organisation. Part of this phase of **ISO 14001:2015** also requires that the needs and expectations of interested parties are fully understood.

The scope of the EMS must be determined and documented, and made available to interested parties. The scope needs to include various issues such as compliance obligations, activities, products and services that the organisation undertakes/offers, in addition to what it can control and influence.

ISO 14001 requires that an organisation establish, implement, maintain and continually improve an **ISO-14001**-compliant environmental management system.

> **DEFINITIONS**
>
> **MANAGEMENT SYSTEM**
>
> Set of interrelated or interlacing elements of an organisation to establish policies and objectives and processes to achieve those objectives.
>
> **ENVIRONMENTAL CONDITIONS**
>
> State or characteristic of the environment at a set point in time.
>
> **ENVIRONMENTAL MANAGEMENT SYSTEM**
>
> Part of the management system used to manage environmental aspects, fulfil compliance obligations and address risk and opportunities.
>
> **PROCESS**
>
> Set of interrelated or interacting activities which transform inputs into outputs. A process may or may not be documented.

Leadership

Top management must demonstrate leadership and commitment to the EMS. Examples of how this can be achieved are by being held accountable for the EMS, ensuring that the EMS is fully integrated into the organisation and promoting continual improvement.

Top management have defined duties as stated in a formal EMS which include responsibility for appropriate resource allocation and the ongoing input and review of the EMS.

Environmental roles and responsibilities must be assigned and communicated by top management in order to comply with **ISO 14001**. Every employee should be made aware of their responsibility in achieving compliance with the policy and specific requirements of the EMS that are relevant to them. Responsibilities can be stated in numerous ways and may be integrated with other job roles in job descriptions and/or a section in the environmental manual.

It should be noted that organisational structures are different in different organisations; however, a sample list of responsibilities is provided in the following table:

Sample environmental responsibilities

Example Environmental Responsibilities	Responsibility
Identify overall direction of the EMS.	Chief Executive/Managing Director.
Design policy.	Chief Executive/Managing Director/ Environmental Manager.
Identify environmental objectives, targets and programmes.	Departmental Managers.
Monitor EMS performance.	Environmental Manager.
Identify training needs/Retain training records.	Environmental Manager/Human Resources Manager.
Track cost associated with the EMS.	Finance.
Identify customer requirements.	Sales and marketing staff.
Compliance with procedures.	All staff.
Undertaking audits.	Audit team.

By determining key on-site issues, it is possible to identify the required roles and responsibilities to ensure effective control. Important areas where responsibilities should be defined are:

- Environmental management programmes (action plans).
- Legislative requirements.
- Control of significant environmental impacts.
- Current responsibilities for environmental management or other management systems (e.g. quality).

Environmental Policy

After significant aspects have been identified, an environmental policy statement can be written. The purpose of a policy statement is to document the environmental intentions and principles of an organisation and to provide a framework for setting objectives and targets. An environmental policy statement is usually of about one page in length and forms a useful marketing tool for the organisation.

2.2 The Key Features and Appropriate Content of an Effective EMS

Senior management commitment should be clearly identified by the most senior person in the organisation signing the policy.

ISO 14001 states the following principles that a policy must comply with; it must:

- Be appropriate to the purpose and context of the organisation, such as the nature, scale and environmental impacts of its activities, products or services.
- Include a commitment to continual improvement and to the protection of the environment (including prevention of pollution and other relevant specific commitments).
- Include a commitment to fulfil its compliance obligations (both legal and other requirements).
- Provide the framework for setting environmental objectives - the policy must identify general aims on how the organisation is to improve; these are backed up by more specific objectives.
- Be documented, communicated within the organisation and available to interested parties.

Planning

Actions to Address Risks and Opportunities

Processes to establish, implement and maintain must be developed to meet the requirements of the planning section of the standard. Whilst planning, the organisation must consider numerous factors such as context of the organisation, the needs and expectations of interested parties and the scope of the system. It also requires that an organisation determines the risks and opportunities related to its environmental aspects, compliance obligations and other issues. The scope of the EMS must include potential emergency situations.

Documented information must be maintained stating the organisation's risks and opportunities that are required to be addressed, in addition to the processes needed to ensure that the requirements of the general section are met as planned.

Environmental Aspects

The organisation must determine aspects and impacts that it can control and influence. This is a key part of **ISO 14001**.

ISO 14001 states that identification of environmental aspects should be carried out for activities, products or services that can be controlled by an organisation, or the activities, products or services that an organisation is expected to have reasonable influence over. It allows identification of aspects that have a significant environmental impact.

ISO 14001:2015 states that a life-cycle perspective should be considered when determining environmental aspects and impacts. This means identification of aspects and impacts from the cradle to the grave (from the development of raw materials all the way to the final disposal of the product). We will cover this concept in more detail in Element 3.

The way in which significant aspects are identified and assessed must be documented. This details responsibilities and arrangements for identifying aspects, determining significance and periodically updating the information.

The Key Features and Appropriate Content of an Effective EMS | 2.2

> **TOPIC FOCUS**
>
> Various criteria are considered to determine '**significant**' impacts during normal, abnormal and emergency operating conditions:
>
> - Scale, severity and duration of the impact.
> - Likelihood of the event occurring.
> - Sensitivity of the receiving medium, e.g. presence of protected species or an already heavily polluted watercourse.
> - Risk of prosecution.
> - Cost of avoiding the impact.
> - Cost of any clean-up.
> - Adverse reaction of the local community and effect on public image.

Compliance Obligations

An organisation is required to have access to documented compliance obligations related to its environmental aspects and determine how these apply. They must also take into account compliance obligations when developing, maintaining and continually improving the EMS.

> **DEFINITION**
>
> **COMPLIANCE OBLIGATIONS**
>
> Legal requirements that an organisation must comply with or another requirement that an organisation has chosen to or must comply with (such as agreements with non-governmental organisations or public authorities, and compliance with voluntary codes of practice).

Planning Action

An organisation must plan to take action to address its significant aspects, compliance obligations and risks and opportunities. It must also plan to integrate the actions into its EMS processes. Technological, financial and business requirements should be considered in planning.

Environmental Objectives

> **DEFINITION**
>
> **OBJECTIVE**
>
> Result to be achieved. An objective may be strategic, operational or tactical. It may apply to different disciplines and different levels. An objective may be expressed in other ways such as an intended outcome, as an environmental objective or by use of other words (aim, goal or target).

To ensure that the commitments stated in the policy are met, objectives need to be developed. Objectives change the nature of the EMS from identifying areas of concern (aspects and impacts) to improving them.

An important part of an EMS is the commitment to and attainment of continual improvement. It is the objectives that help provide evidence for this improvement.

ISO 14001:2015 is vague as to the types of objectives that must be set. There is just a requirement to set objectives and no mention is made as to specific mandatory requirements.

One method, however, is that organisations initially develop high level environmental objectives which are broad areas of improvement that are normally not quantified. Lower level objectives (sometimes known as targets) are more detailed and are linked to higher level objectives. Usually such lower level objectives are SMART:

- **S**pecific.
- **M**easurable.
- **A**chievable.
- **R**ealistic.
- **T**ime-bound.

Action plans tend to be very low level objectives - they identify how the requirements of the linked higher level objectives will be met. As a minimum they tend to consider:

- **Tasks** - to be completed to meet a linked objective.
- **Timescale** - by when a task will be completed.
- **Responsibility** - who will be responsible for ensuring that the task is completed.

In practice, they are often limited to addressing significant aspects but they may consider other issues for business or financial reasons.

Organisations will often develop **environmental performance indicators** (a form of key performance indicator) that measure reductions in significant environmental impacts. These are routinely used internally to measure performance but can be externally validated.

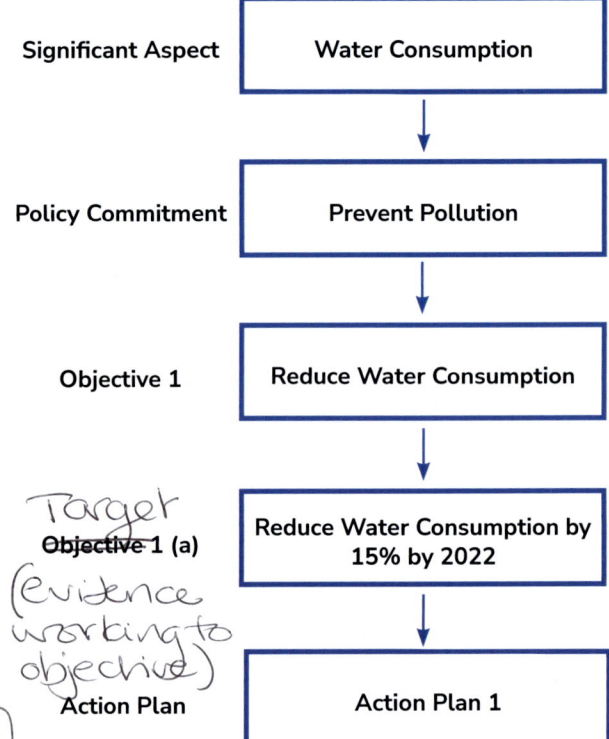

Target (evidence working to objective)

Environmental objectives example

Methods of Selection of Objectives

There is no fixed way of selecting areas where objectives should be set. Determining the type of objective will involve carefully balancing:

- the extent that an impact is an issue to a stakeholder;
- the influence of the stakeholders;
- how well the impact is already being managed; and
- the resources available.

For many organisations the need for improvement is based on what the law requires. In such situations, what the law requires may be the maximum that the organisation can cope with.

The **ISO 14001** standard states that environmental objectives must be documented and be:

- Consistent with the environmental policy.
- Measurable (the standard states if practicable).
- Monitored.
- Communicated.
- Updated as appropriate.

Whichever method is used to set objectives, the **ISO 14001** standard states that when planning objectives an organisation must determine:

- What needs to be done.
- What resources will be needed.
- Who will hold responsibility for the objective.
- When the objective is due for completion.
- How results will be evaluated.

Support

Resources

The organisation must understand and provide the resources required for the implementation, maintenance and continual improvement of the EMS.

Competence

> **DEFINITION**
>
> **COMPETENCE**
>
> Ability to apply knowledge and skills to achieve intended results.

Various requirements for competence are present in this part of the standard such as:

- Understanding the required competence of those whose work could affect environmental performance or compliance obligations.
- Ensuring that such persons are competent (education, training or experience).
- Understanding training needs associated with the environmental aspects and EMS.
- Taking action to gain necessary competence and evaluate such actions.
- Retaining documented information as evidence of competence.

Awareness

Those who carry out work under the organisation's control must be aware of the environmental policy, significant environmental aspects and impacts, their contribution to the effectiveness of the EMS and implications of not complying with the EMS.

Communication

Having effective communication structures is important to:

- Motivate the workforce.
- Explain the environmental policy (both internally and externally) and how it relates to the overall vision/strategy of the organisation.
- Ensure understanding of roles and responsibilities.
- Demonstrate management commitment.
- Monitor performance.
- Identify potential system improvements.

2.2 The Key Features and Appropriate Content of an Effective EMS

ISO 14001 requires that an organisation has processes that cover:

- Internal communications between the various levels and functions of the organisation.
- External communication of information relevant to the EMS as stated in the organisation's communication process and required by compliance obligations.

Such processes need to cover:

- What will be communicated.
- When to communicate.
- With whom to communicate and the methods of communication.

The organisation must also consider processes for external communication on its significant environmental aspects and record its decisions.

Documented Information

Specific requirements are present in the standard for various information to be documented, in addition to information deemed to be necessary for the EMS by the organisation. There are also documented information requirements for:

- Creating and updating: documented information: should be able to be identified (e.g. date, title, reference number), a correct format should be used (language, software version, etc.) and reviewed and approved for suitability and adequacy.
- Document control: documents must be controlled to ensure availability and adequate protection (loss of integrity). More specifically, an organisation must consider distribution, storage, preservation, control of changes and retention.

Operation

Certain activities and operations must be controlled within the EMS. The standard states that controls can be both procedural and engineering and may follow a hierarchy of elimination, substitution and administrative.

Consistent with a life-cycle perspective the organisation must:

- Develop controls to ensure environmental requirements are considered at each life-cycle stage during the development process of products and services.
- Determine what requirements are needed for the procurement of products and services.
- Communicate relevant environmentally-related information to external providers (e.g. contractors).
- Consider whether to provide information regarding significant impacts associated with transport, delivery, end-of-life treatment and end-of-life disposal of products and services.

Significant aspects should be controlled. This would be carried out by considering the activities that cause these aspects and what type of control is required to manage or minimise impacts. A draft procedure could then be developed, which should be trialled and amended as required, prior to issuing of the final version.

Emergency Preparedness and Response

Organisations must develop and maintain documented processes to identify and respond to accidents and emergencies and to prevent or reduce environmental impacts that are associated with them. The processes must be reviewed and revised on a regular basis, particularly after an accident has occurred. They must also be tested.

Common accidents that have an environmental impact include:

- Fires.
- Floods.
- Wastewater releases.
- Air releases.

The number and type of procedures that are required to be developed depend on the type and complexity of an organisation; e.g. a large chemical company will need a relatively complex emergency plan, whereas an office would require a few simple procedures.

> **TOPIC FOCUS**
>
> **Emergency preparedness and response plans** may include the following elements:
>
> - On-site emergency response teams and equipment.
> - Key personnel duties, responsibilities and contact details.
> - Interrelationship with, and contact details for, off-site emergency services.
> - Internal and external communication plans.
> - Training arrangements and practice drills.
> - Detailed response measures for each type of emergency incident - including personnel response and equipment needs.
> - Inventories, locations, method of storage and potential effects on the environment of the full range of chemicals held on the site.

Performance Evaluation

Scope and Purpose of Auditing Environmental Management Systems

> **DEFINITION**
>
> **ENVIRONMENTAL AUDIT**
>
> *"Systematic, independent and documented process for obtaining audit evidence and evaluating it objectively to determine the extent to which the audit criteria are fulfilled."*
>
> ISO 14001:2015

An audit will usually only look at a sample of the processes and activities that are taking place within an organisation, and the documentation controlling them, so it is essential that the process is carried out in a systematic way. This helps ensure that the results of the audit are objective and that different auditors achieve similar results. Objective results are based on evidence, not hearsay, so it is essential that an auditor follows a trail of evidence to a final conclusion. For example, if, on inspecting emissions-testing results, the first is found to be over a prescribed limit but the auditor is told this is exceptional and unusual, he must look further at other results to ascertain for himself whether this is the case or not.

The audit criteria must be defined before the audit process starts, as these are the terms against which the audit will be judged. For example, an EMS audit being undertaken with a view to gaining certification to the **ISO 14001** standard will be to confirm whether or not the organisation's EMS conforms to the requirements of the **ISO 14001** standard.

The scope of the audit will be defined by the scope of the EMS, as this is what is being audited; the purpose will be to identify that the EMS complies with the standard.

2.2 The Key Features and Appropriate Content of an Effective EMS

Distinction Between Audits and Inspections

There is often confusion between inspections and audits and it is important to understand the differences between them:

- Inspections are an assessment of what is there at the time an inspection takes place. They are essentially looking for signs of failure, e.g. leaks in pipes, broken lights, signs of damage, etc.
- Audits are a pre-planned, systematic and objective assessment of a situation against a given set of criteria. For example, clause 5.2 of **ISO 14001:2015** states:

    "Top management shall establish, implement and maintain an environmental policy that, within the defined scope of its environmental management system ... is appropriate to the purpose and context of the organisation."

    The auditor would be looking for evidence to establish that a policy is in place, has been agreed to by top management and is appropriate to the nature and scale of the organisation.

Internal Audit

ISO 14001 states that audit programmes should cover:

- The activities and areas to be considered in audits.
- The frequency of audits.
- The responsibilities associated with managing and conducting audits.
- The communication of audit results.
- Auditor competence.
- How audits are to be conducted.

The organisation must define audit criteria for each audit, select auditors who are objective and impartial and ensure that results are communicated to management. Documented evidence of the implementation of the audit programme must be retained.

The scope of an EMS audit is vast, but it basically seeks to determine whether the EMS conforms to **ISO 14001**, legal issues and procedures. It should also provide information to show if the EMS is properly implemented and maintained. The audit report is required to provide information to management on the results of audits.

The internal audit plays an important role within the EMS and therefore is an important component of the certification assessment.

Internal audits of the EMS are often performed by personnel from within the organisation, although external auditors may be used. In any event, persons conducting the audit should be competent and have no responsibility for the activity being audited.

Environmental Auditing

Pre-Audit Preparations

Before an audit takes place, some preparation is required by both the audit team and the auditee (the person or organisation being audited). An important requirement is the selection by the lead auditor of suitable auditors. The main requirement is to ensure that necessary auditor competencies are available so that the audit objectives can be met. The following factors should be considered when determining the size and members of the team:

- Objectives, scope, criteria and duration of the audit.
- The competencies of the audit team to meet the objectives.
- Legal, contractual and certification body requirements.
- The requirement to ensure that all auditors are independent.
- The language in which the audit will be undertaken.

The Key Features and Appropriate Content of an Effective EMS

Other key factors in audit preparation include:

- The scope of the audit must be defined, e.g. a specific factory or office location, or a production process and the EMS surrounding that process.
- There may be logistical issues to arrange, such as:
    - Transport around a large site.
    - Correct Personal Protective Equipment (PPE).
    - Sufficient employees to accompany the auditors.
    - Office space, both for the auditors to work in and for opening and closing meetings.

Information-Gathering

An essential starting point in any audit process is the gathering of existing information that will inform the direction and focus of the audit. This will help ensure that the auditor is as familiar as possible with the organisation's activities and highlight particular areas or questions that the auditor needs to focus on. Much of the information-gathering will be based on a review of documentation, especially:

- Site plans.
- The results of previous audits.
- Records of emissions monitoring (versus legal limits).
- Accident/incident reports.
- Enforcement notices.

Notifications and Interviews

Key personnel must be informed if they are to be involved in the audit, as it is important that they are available and they may have their own preparation to complete before the audit. Interviews should be conducted in a reasonably formal manner without being confrontational. Many people feel uncomfortable being interviewed and tend to say as little as possible. The skill of the interviewer is in getting them to talk openly. This helps ensure that both positive and negative issues are aired and the evidence can be identified so that improvements can be made.

Responsibility for Audits

Ultimate responsibility for ensuring that auditing is carried out will rest with senior management. Internal audits should normally be carried out by those with no line-management responsibilities for the specific activities being audited. A team approach is often used, consisting of managers as well as workers. The key thing is that the auditors are competent and responsible. External audits will be required for **ISO 14001** certification.

It is important that the results of the audits are reported to management.

Advantages and Disadvantages of External and Internal Audits

There are advantages and disadvantages to both internal audits (first-party audits) and external audits (second- and third-party audits).

2.2 The Key Features and Appropriate Content of an Effective EMS

> **DEFINITIONS**
>
> **FIRST-PARTY AUDIT**
>
> Often referred to as an internal audit; usually undertaken by the organisation's own auditors.
>
> **SECOND-PARTY AUDIT**
>
> Undertaken by another organisation, often a consultancy, or a team of auditors from another organisation.
>
> **THIRD-PARTY AUDIT**
>
> An independent, formal audit; usually undertaken by an accredited certification body.

	Internal Audit	External Audit
Advantages	<ul><li>Reduced costs.</li><li>Auditor probably has better understanding of the business.</li><li>Interviewees may relax more with someone they know, or who at least is from within the organisation.</li><li>More flexibility on times for audits and potential to change if needed.</li></ul>	<ul><li>Objective auditor with no preconceived perceptions.</li><li>Auditor likely to have experience of other similar processes/industries, so able to bring a comparative view.</li><li>Seen by external parties (customers, suppliers and regulators) as more objective and carries more weight.</li><li>Once audit is booked, it is a set deadline by which to achieve objectives.</li><li>Auditor less likely to be unduly influenced by senior managers not to highlight poor areas.</li></ul>
Disadvantages	<ul><li>Auditor may be 'blind' to some problems, as they are 'situation normal'.</li><li>Seen by external parties (customers, suppliers, etc.) as less objective.</li><li>Auditor possibly more open to pressure from senior management not to highlight negative points.</li><li>Flexibility may lead to audits continually being put off.</li><li>Time required for employees to conduct audits and also attend training.</li></ul>	<ul><li>Costs increased.</li><li>Less flexibility on times; once audit is booked it is likely to be charged for, so greater pressure to undertake even if not quite ready.</li><li>Auditor may not fully understand some of the issues involved.</li><li>Interviewees may not feel they want to discuss problem areas with an external person.</li></ul>

Monitoring, Measurement, Analysis and Evaluation

An organisation's environmental performance must be monitored, measured, analysed and evaluated. An organisation must determine:

- What should be monitored and measured.
- The methods for monitoring, measuring, analysing and evaluating.
- What criteria the organisation's performance is compared against.
- The frequency of monitoring.
- When monitoring results should be analysed and evaluated.

The Key Features and Appropriate Content of an Effective EMS | 2.2

The organisation is also required to communicate performance both internally and externally (this should be identified in the communication process). Documented evidence of monitoring must be kept.

Processes must be developed, implemented and maintained to demonstrate that the organisation has fulfilled its compliance obligations.

Such evaluation processes must include the frequency of compliance evaluation, any action required and maintenance of knowledge and understanding of organisational compliance status.

Documented evidence of the evaluation of compliance must be retained by the organisation.

Active Monitoring Measures

'Active' (sometimes called 'proactive') simply means measuring progress towards targets before anything has gone wrong and possibly looking for trends that identify potential problems in the future.

> **TOPIC FOCUS**
>
> **Active monitoring measures** could include:
>
> - Emissions - monitoring atmospheric or water emissions (and comparison with permit conditions).
> - Effluent - measuring the flow rate of a discharge to water.
> - Waste - monitoring production of waste over time.
> - Energy and water - monitoring use over time can provide valuable information on the efficiency of machinery and the presence of leaks for water.
> - Mass balance calculations (measuring materials in and out of a process, checking for leaks).
> - Key Performance Indicators (KPIs) (see below) for all relevant parties, such as senior managers, procurement department, and suppliers involved in the environmental management of the organisation.
> - Inspection of plant and premises.

> **DEFINITION**
>
> **KEY PERFORMANCE INDICATORS**
>
> Help define and measure performance towards an objective. They are a key part of a measurable objective.
>
> Example KPIs could include emission of greenhouse gases or metal emissions to water. To be measurable the target would need to be quantified.

Use of Environmental Inspections and Tours

Inspections and tours can be used as tools to gather evidence of environmental performance. An environmental inspection should:

- Be a formal, organised walk-round of a workplace, or section of a workplace.
- Identify uncontrolled environmental impacts and non-compliances.
- Ensure that suitable controls are put in place.

A checklist is typically used. Minor defects may be rectified at the time but others may need finance and other resources to resolve them.

Inspections can be grouped into three main types:

- General environmental inspections - undertaken by managers, environmental representatives, or other suitably trained members of staff.
- Statutory inspections - required by legislation.

2.2 The Key Features and Appropriate Content of an Effective EMS

- Compliance inspections:
    - To compare against stated performance standards, which may include legal compliance and may also form part of an audit.
    - Undertaken by staff from an internal control department, or by an enforcement agency.

A tour, as opposed to an inspection:

- Usually follows a predetermined route through the area or workshop.
- Typically lasts only 15 minutes or so.
- May be conducted at weekly intervals to ensure that standards are maintained.

Factors Governing Frequency

The frequency of the inspections will depend on issues such as:

- The purpose of the inspection.
- Any frequency imposed by regulations, such as discharge consent or environmental permits.
- The level of risk to the environment.
- Conditions found at the last inspection.

Competence of Inspector

It is important that the person carrying out the inspection is competent to do so and, in some cases, it may be necessary for several people to carry out the inspection together to cover all the areas of expertise required. For example, in a large production process, such as cement manufacture, it may be necessary for a production manager or supervisor to be accompanied by the environmental manager, as one is likely to understand the process in depth and the other will have greater understanding of the legal obligations imposed by an environmental permit.

Anyone carrying out an environmental inspection should have, as a minimum:

- An understanding of the tools of workplace inspections, their advantages and disadvantages, and how to use them.
- An understanding of the process or activity being inspected.
- Knowledge of the potential environmental impacts from the process or activity.
- Knowledge of the standards that are acceptable.
- A basic report-writing ability, or ability to use a checklist.

Use of Checklists

Preparing a checklist will contribute to the success of an inspection.

A checklist:

- Provides a reference point to ensure the inspection remains focused.
- Can be deviated from, e.g. to investigate an unexpected problem or defect, but provides a 'bookmark' for the inspector to continue the inspection from where he/she left off.
- Aids note-taking and provides an easy way to follow up on issues found during an inspection that could not be followed up immediately.

There is no set format for a checklist, as this will vary with the activity being inspected and the individual preference of the inspector.

The Key Features and Appropriate Content of an Effective EMS | 2.2

> **TOPIC FOCUS**
>
> Examples of the areas that may be covered in a general **checklist** include:
>
> - Site drainage - marked with type of drain and direction of flow.
> - Fugitive emissions - unplanned emissions.
> - Waste - correct type of containers, correct waste in containers, segregated and secure.
> - Discharges from interceptors - visible signs of oil.
> - Noise - unusual noise levels.
> - Signs - are they clear and correct?
> - Signs of leaks from machinery or other spills.
> - Are unattended machines left running?

Allocation of Responsibilities

Some faults will need specialist knowledge to resolve, and others will require the allocation of finance and resources. Responsibilities for corrective and preventive actions will therefore need to be allocated.

Priorities for Action

Priorities for dealing with the issues identified will need to be set. These priorities should be based on the level of risk to the environment, even though some of the actions will take longer to complete owing to the level of resources required. Issues affecting legal compliance must be allocated a high priority, as these are likely to not only have significant environmental impact but also a significant impact on the business, should enforcement action be taken.

Reactive Monitoring Measures

Reactive monitoring is effectively the monitoring of organisational failures with a view to preventing a recurrence of the failure. So, it is a reaction to an undesirable event that has already occurred. Lessons need to be learnt and both corrective and preventive actions taken. 'Reactive monitoring measures' are simply data arising from these reactive monitoring techniques.

> **TOPIC FOCUS**
>
> **Reactive monitoring measures** could include:
>
> - **Near-misses** - unplanned, unwanted events that have the potential to cause environmental damage. Oil getting into a drain system, for instance, but being caught before it reaches controlled waters, would be a near-miss.
> - **Complaints/suggestions** from neighbours and/or the workforce - often an early sign that something has changed in the process, creating a potential problem. This may be noise, dust, odour, etc.
> - **Waste-stream monitoring** - the quantity and type of waste produced. Increases in waste going to final disposal, for example, may indicate a deterioration in waste management control.
> - **Energy** - measuring energy usage in different buildings or different parts of a building will identify areas of excessive energy use.
> - **Water** - recording water usage will identify trends in water consumption.
> - **Enforcement action** - a major signal that things are not right. Enforcement action shows that not only are there potential deficiencies within the main process but also with the way active monitoring is carried out, or the way the results of that active monitoring are dealt with.

2.2 The Key Features and Appropriate Content of an Effective EMS

Review of Environmental Performance

The purpose of reviewing environmental performance is to analyse the data gathered through the monitoring techniques discussed above and to make decisions on whether that performance is acceptable. The review can then question whether the organisation:

- Is achieving its environmental objectives.
- Is implementing effective environmental controls.
- Is being effective in its training, communication and consultation with employees and other stakeholders.
- Learns lessons from environmental incidents and implements effective corrective and preventive actions.
- Is legally compliant.
- Reduces the risk of causing environmental damage.

Identifying strengths and weaknesses, and highlighting where standards have dropped or changes have occurred, will allow the review process to maintain the momentum in the EMS and to continually improve and manage change.

Gathering Information to Review Environmental Performance

> **TOPIC FOCUS**
>
> Various **sources of information** can be used to provide data for both the initial and the management environmental review:
>
> - Incident data - often provides information on effectiveness of procedures and training currently in place. Although a failure has occurred, there is an opportunity to learn from it and improve in the future.
> - Inspections - opportunities to identify potential incidents before they occur and rectify problems early.
> - Control and monitoring of emissions - ensures compliance with any legal requirements on emission levels and allows trends to be analysed to identify any potential future problems.
> - Energy/raw material management - changes in volumes of raw materials and energy per unit of production may be an early indication of failures developing in the system.
> - Waste management - increases in waste generated per unit of production act in a similar way to energy and raw materials.
> - Surveys, tours and sampling - as for inspections.
> - Quality assurance reports - provide similar data to raw materials and energy usage and waste produced. Together they will provide an indication of changes in the process.
> - Audits - highlight areas for improvements in either the procedures or the implementation of those procedures.
> - Monitoring data/records/reports - as for inspections and control and monitoring of emissions; will usually support the findings of inspections, audits, etc.
> - Complaints - a valuable independent source of data that can lead to indications of improvements required in the processes, or the procedures for monitoring.

> **DEFINITIONS**
>
> **MONITORING**
>
> Methods used by an organisation to measure how effectively policies are being implemented, how well they are controlling environmental risks and how the culture of the organisation is developing with regard to environmental protection.
>
> **REVIEW**
>
> Analysis of the data gathered through the monitoring processes, allowing an organisation to judge whether environmental risks are adequately controlled.
>
> **AUDIT**
>
> An objective and systematic assessment of the organisation's EMS in order to determine if systems exist, are adequate, and are used.

Investigating Environmental Incidents and Reporting Requirements Internally and Externally

Environmental incidents would normally be reported internally (within the organisation) and may also need to be reported externally (e.g. to enforcement agencies and insurance companies). Management may investigate environmental incidents, or at least take an active interest in the findings of investigation reports, and these may all contribute to the management review.

Reporting on Environmental Performance

Review findings can also be incorporated into environmental performance reports, both internally and externally to, for example, shareholders and other stakeholders.

Role of Boards, Chief Executive/Managing Director and Senior Managers

Selected members of the senior management team should perform the periodic review. This ensures Board involvement, giving the necessary authority and weight to the process.

Feeding into Action and Development Plans as Part of Continual Improvement

Each periodic review should naturally output decisions and actions needed for improvement of the EMS. If successfully implemented, this should lead to continual improvement over time.

Management Review

The **ISO 14001** standard identifies that an organisation's top management must review the EMS to ensure that it is suitable, adequate and effective. The review should also be completed at specified intervals and the inputs and outputs should be communicated to relevant staff. The management review must be developed to assess the needs for EMS improvements. Review frequency is not identified in **ISO 14001**; however, most organisations undertake a management review on an annual basis. This is a formal review but informal reviews can also be undertaken between formal management reviews.

The standard states that the management review must include:

- Status of actions from previous reviews.
- Changes in significant aspects.
- Internal and external issues.
- The extent to which objectives have been reached.
- Information on the environmental performance of the organisation, such as trends in monitoring results, fulfilment of compliance obligations and audit results.

- How adequate resources are.
- Communications from interested parties (this includes complaints).
- Areas where there are opportunities for continual improvement.

The output of the management review will include:

- Whether the EMS is still suitable, adequate and effective.
- Decisions that are linked to continual improvement.
- Action when environmental objectives have not been met.
- Areas where the EMS could be integrated with other business processes.
- Implications for the strategic direction of the organisation.

Documented evidence of the results of the management review must be kept.

Improvement

An organisation must determine where improvement opportunities exist and implement actions to ensure achievement of EMS outcomes.

Nonconformity

> **DEFINITIONS**
>
> **CONFORMITY**
>
> Fulfilment of a requirement.
>
> **NONCONFORMITY**
>
> Non-fulfilment of a requirement.

If a nonconformity occurs, an organisation must take necessary action to ensure that it is controlled and rectified. It must also react to the consequences of the nonconformity such as mitigating the adverse environmental impacts that may occur.

Action is also required to eliminate the cause of the nonconformity to ensure that it does not happen again. This is to be achieved by undertaking a review of the nonconformity, understanding the causes and to determine whether a similar nonconformity exists or could occur. Any corrective action implemented must be reviewed for its effectiveness.

Documentary evidence must be retained as to the nature of nonconformities, action undertaken to correct them and the results of the corrective actions.

Continual Improvement

> **DEFINITION**
>
> **CONTINUAL IMPROVEMENT**
>
> Recurring activity to enhance performance. Enhancing performance is related to the use of the EMS to improve environmental performance consistent with the environmental policy.

The organisation is required to ensure continual improvement of the EMS's suitability, adequacy and effectiveness to enhance environmental performance.

The Key Features and Appropriate Content of an Effective EMS | 2.2

Eco-Management and Audit Scheme (EMAS)

In addition to the international EMS standard **ISO 14001**, a European-based standard - the **Eco-Management and Audit Scheme (EMAS)** - has been developed (under **EC Regulation 1221/2009** as amended by **EU Regulation 2017/1505**). EMAS is a European regulation that enables industries to voluntarily implement formal environmental management systems.

EMAS shares a common core framework with **ISO 14001** and so provides an organisation with a structured approach for identifying, evaluating, managing and improving its environmental performance. Although similar in content, EMAS has a number of important differences from **ISO 14001**. These include:

- A formal environmental review must be documented as a precursor to establishing the system.
- An independently verified Environmental Statement must be prepared, which sets out key information for the public about the organisation's impacts and actions.
- An open dialogue must be established with the public and other interested parties.
- EMAS uses stronger and more specific language about legal compliance than **ISO 14001** (for example, the organisation must have identified and know the implications to them of environmental legal requirements). Breaches of legislation may result in EMAS registration being withdrawn.
- EMAS is site-based whereas **ISO 14001** can be organisation-wide (although there is scope for EMAS multi-site verification for listed industry sectors such as office administration and support and management consultancy).
- EMAS has a three-year audit cycle - there is no specific audit cycle set in **ISO 14001**.

However, the key difference between EMAS and **ISO 14001** is that EMAS is a European, rather than international standard. There are many fewer EMAS registrations worldwide - mostly held by larger industrial companies, to which EMAS is best suited.

STUDY QUESTIONS

2. What is an environmental policy?
3. List the seven key stages of the **ISO 14001** environmental management system standard.
4. Who should prepare and endorse the environmental policy?
5. List any three of the main inputs into a management review of an environmental management system.
6. What factors should be considered when deciding how frequently environmental inspections should be held?
7. What should be the minimum requirements for someone carrying out an environmental inspection?
8. Explain the difference between an environmental audit and an environmental inspection.

(Suggested Answers are at the end.)

2.3 Benefits and Limitations of Introducing a Formal EMS into the Workplace

Benefits and Limitations of Introducing a Formal EMS into the Workplace

IN THIS SECTION...
- There are benefits associated with introducing **a formal EMS** into the workplace.
- There are also limitations when introducing **a formal EMS** into the workplace.

Benefits of Introducing a Formal EMS into an Organisation

Many people argue that introducing and operating an EMS in an organisation is free of charge. This is not to say that there are no costs involved, but that the benefits gained from the implementation and operation of an EMS far outweigh those costs and therefore there is a net benefit. It is clear that there are many benefits associated with the implementation of an EMS and certification to the **ISO 14001** standard or other formal standards.

Accreditation and Certification

Formal environmental management systems can be externally **certified**. Certification bodies are external bodies which certify an organisation to the requirements of EMS standards such as ISO 14001 by undertaking audits of the organisation's EMS against the requirements of the standard. Examples include the British Standards Institution (BSI) and TÜV Rheinland.

Accreditation is the process by which certification bodies are checked to ensure competence. Accreditation bodies provide an accreditation service for certification bodies, ensuring that they provide a high quality certification service. For example, the accreditation body for ISO 14001 certification bodies in the UK is UKAS. Accreditation of certification bodies is generally voluntary.

> **TOPIC FOCUS**
>
> **Benefits of Introducing ISO 14001**
>
> - Increased compliance with legislative requirements:
>   - An organisation must identify relevant compliance obligations such as legislation, and this surely is the first step to compliance.
>   - Written procedures are in place so that employees understand how they can comply with the legislation.
>   - The profile of legislation is raised within the management team and highlights the potential penalties for non-compliance.
>   - Where procedures are written in such a manner as to ensure activities are carried out in compliance with legislation, compliance is more likely to be assured.
> - Competitive edge over non-certified businesses.
>
>   Many organisations now require suppliers to provide evidence of their commitment to improving their environmental performance and reducing their impact on the environment. A certified management system is excellent evidence for this and, in many cases (such as many local government contracts), it is a prerequisite for tendering.
>
> - Improved management of environmental risk.
>
>   The compilation of the Aspects and Impacts Register (see later) ensures that the significant impacts are identified and, once identified, they are more likely to be controlled sufficiently to prevent unauthorised pollution incidents from occurring.
>
> (Continued)

Benefits and Limitations of Introducing a Formal EMS into the Workplace — 2.3

> **TOPIC FOCUS**
>
> - Increased credibility that comes from independent assessment.
>
>   While it is not a requirement of the standard to undergo external certification, there are significant advantages, as the organisation potentially gains considerably increased credibility in the marketplace.
>
> - Savings from reduced non-compliance with environmental regulations:
>   - There may be reduced costs for permits where there is better environmental performance.
>   - The risk of incurring costs, such as fines and penalties for non-compliance, is reduced.
>
> - Heightened employee, shareholder and supply-chain satisfaction and morale.
>
>   Investors and employees want to be associated with positive environmental images.
>
> - Meeting modern environmental ethics.
>
>   The ethical investment market is growing steadily - from a niche marketplace a few years ago to now being a mainstream opportunity.
>
> - Streamlining and reducing environmental assessments and audits:
>   - Integration of quality, health and safety and environmental management systems can result in more streamlined assessments and audits.
>   - Audits will be more systematic and planned (instead of haphazard) and demonstrate continual improvement.
>
> - Increased resource productivity.
>
>   It has been shown that those companies with certified management systems are also some of the best performing companies in the marketplace, as these are also the better managed companies.

Limitations of Introducing a Formal EMS into an Organisation

While there are many benefits, as discussed above, to implementing a management system certified to a formal standard, it is not without some limitations.

> **TOPIC FOCUS**
>
> **Limitations of Introducing ISO 14001**
>
> - Prescriptive environmental performance levels are not included within the standard.
>
>   Organisations must set what they consider to be relevant performance levels and it is not easy to compare one organisation in a particular industry with others in the same industry.
>
> - Improvements in environmental performance can be negligible.
>   - For some organisations, especially small, low-impact ones, there may be very little room for them to improve their impact, especially in ways that will also have a significant positive impact on the performance of the company.
>   - Some organisations also find that as the system matures, it is difficult to satisfy the requirement to continually improve their performance.
>   - This may lead to an organisation focusing on trivial and insignificant issues that have little impact on the environment but consume significant resources in time and money.
>   - This may lead to an organisation focusing on trivial and insignificant issues that have little impact on the environment but consume significant resources in time and money.
>
> (Continued)

2.3 Benefits and Limitations of Introducing a Formal EMS into the Workplace

TOPIC FOCUS

- Lack of public reporting, unlike other internationally recognised management systems.

    Unlike the **Eco-Management and Audit Scheme (EMAS)**, **ISO 14001** does not require an organisation to publicly report its environmental performance.

- The **14001** standard can be self-certified but is much more credible if certified by an external body. There is, however, often considerable variation in the approach that individual auditors take towards certifying organisations. This is often dependent upon the individual's knowledge and experience of the particular industry being audited, which could range from chemical works and nuclear power stations to offices and leisure facilities.

- Implementing an EMS may have high cost implications for small- and medium-sized enterprises:

    - Implementing an EMS is not cheap.

    - Typically, implementation takes six to eight months and will require the services of an experienced consultant to at least direct progress towards certification.

    - There is also the cost of external certification bodies' initial and re-certification audits and of the more frequent surveillance audits, between certification visits.

STUDY QUESTION

9. List three benefits and three limitations associated with the implementation of a certified EMS developed to the **14001** standard.

(Suggested Answer is at the end.)

Summary

This element has dealt with the key features, benefits and limitations of environmental management systems.

In particular, this element has:

- Identified some of the main reasons for implementing an Environmental Management System (EMS):
    - It demonstrates senior management commitment to environmental management.
    - Environmental, health and safety and quality share common management system principles, so the three systems can be integrated into an environmental, health, safety and quality system.
    - It is often required by stakeholders (e.g. customers or regulators).
    - It demonstrates a sense of corporate responsibility.
- Shown how an initial environmental review should cover all current activities and answer the question: "Where are we now?"
- Explained that the environmental policy should:
    - Be appropriate to the nature and scale of the organisation.
    - Include a commitment to continual improvement.
    - Fulfil compliance obligations.
    - Provide a framework for setting and reviewing of objectives.
    - Be communicated to interested parties.
- Described how the organisation's planning should include:
    - Up-to-date information in relation to, and evaluation procedures to determine, significant actual or potential environmental impacts.
    - Procedures for identifying compliance obligations relating to land and buildings, processes and plant, products and services, and environmental performance standards it should achieve.
    - Establishment of environmental objectives.
    - Environmental management programmes.
- Explained how requirements for operation and performance evaluation of an EMS are contained within the clauses of **ISO 14001**.
- Outlined the purpose of an environmental audit to establish that an EMS complies with the requirements of **ISO 14001**.
- Described how both active (regular monitoring of performance standards and systematic inspection of plant and processes) and reactive (data on near-misses and complaints) monitoring techniques should be used to collect information on environmental performance.
- Explained that regular reviews of the EMS should be carried out by senior management, with the aim of implementing continual improvement of performance.

Exam Skills

Question

Scenario

You have been appointed as the environmental adviser to a manufacturing facility with the role of assisting managers in undertaking their environmental management duties. As you are new to the role, you decide to undertake an initial environmental review of the organisation's current environmental performance with the view to developing a formal environmental management system to ISO 14001:2015. As part of the review, you decide to develop a checklist to ensure that the review process is systematic.

Task: Environmental Management System

What are the key internal environmental documents that you will need to check as part of the review that you would include on your checklist? **(8 marks)**

Approaching the Question

Think about the steps you would take to answer the question:

- Read the scenario carefully. With this question, you need to determine what type of document would be viewed during an initial environmental review of a manufacturing facility. 'Manufacturing facility' is quite a broad scenario, so you can use your imagination as to what is actually manufactured.
- Now look at the task – prepare notes (and examples) on:
  - Initial environmental review.
  - Documentation needed for the review.
- Consider the marks available. In this case, there are 8 marks available so you should provide at least 8 internal documents that could be reviewed.
- Read the scenario and task again to make sure you understand them and have a clear understanding of the IER process and documentation that needs reviewing.
- Jot down an outline plan - this might include:
  - Results of past audits, monitoring, complaints records, accident reports, EMS, maintenance, proactive monitoring, staff training records.

Now have a go at the question yourself.

Exam Skills

Example of How the Question Could be Answered

An initial review is carried out to determine the organisation's current position with regard to environmental management. It is undertaken prior to the development of an EMS. The review will cover all areas that are within the scope of the proposed environmental management system. The review checklist is likely to require that the following internal documents are reviewed:

- *The results of previous audits or reviews would be covered. These will identify weaknesses in management and identify areas that may need to be considered in the current review.*
- *Records of monitoring of air emissions, raw-material usage and energy consumption would all need to be considered. Such documents will highlight the level of compliance with emission limits in permits, consents or other documents.*
- *Complaint records will also give an idea of the significant impacts that have occurred and the corrective actions implemented; they may also need to be retained for legal reasons (e.g. requirement of an installation permit).*
- *Any reports from past accidents or incidents would also need to be considered, as they will, again, highlight deficiencies in the management of the company's environmental impacts.*
- *The company's environmental policy, procedures, processes and EMS manual are also likely to be inspected, as these will show the robustness of the management system that the company operates.*
- *Additionally, maintenance logs could be considered in order to check the level of maintenance of key pollution abatement equipment, such as secondary containment for tanks, or air-pollution abatement devices.*
- *The results of proactive monitoring, such as site inspection, could also be considered, as these will form part of the management system of the organisation.*
- *Staff training records will also be considered for staff who could cause a significant environmental impact. Training is important, as it may be an important risk-control measure for the organisation.*

Element 3

Assessing Environmental Aspects and Impacts

Learning Outcomes

- Assess environmental aspects and associated impacts, determining significant aspects and evaluating current controls.

Learning Objectives

Once you've read this element, you'll be able to:

1. Recognise different types of environmental impact.

2. Review and use sources of environmental information.

3. Apply the principles and practice of environmental aspect and impact assessment.

Contents

Reasons for Carrying Out Environmental Aspect and Impact Assessments	**3-3**
Why Identify Environmental Impacts?	3-3
Aims and Objectives of Impact Assessment	3-3
Life-Cycle Analysis (Cradle-to-Grave Concept)	3-6
Types of Environmental Impact	**3-8**
Direct and Indirect Impacts	3-8
Cumulative Impacts	3-8
Positive and Negative Effects	3-8
Nature and Key Sources of Environmental Information	**3-11**
Internal to the Organisation	3-11
External to the Organisation	3-12
Identifying Environmental Aspects and Associated Impacts	**3-13**
Implementing an EMS	3-13
Identifying Environmental Aspects	3-13
Determining Associated Environmental Impacts	3-14
Specific Impacts	3-16
Identifying Receptors at Risk	3-17
Identification of Aspects and Impacts	3-17
Evaluating Impact and Adequacy of Current Controls	3-19
Risk and Opportunities	3-20
Recording Significant Aspects and Impacts	3-21
Reviewing	3-21
Summary	**3-22**
Exam Skills	**3-23**

Reasons for Carrying Out Environmental Aspect and Impact Assessments | 3.1

Statutory assessment

Reasons for Carrying Out Environmental Aspect and Impact Assessments

IN THIS SECTION...

- An Environmental Impact Assessment (EIA) is a tool used to identify environmental impacts, to assess the level of risk from those impacts, and to develop and implement appropriate controls (or mitigation) to reduce the environmental risk.
- Life-Cycle Analysis is a tool used to identify and measure the environmental impact of a product or service throughout its life cycle - a 'cradle-to-grave' approach.

Why Identify Environmental Impacts?

Before any organisation can start to develop an environmental improvement programme it needs to answer the question: "What environmental impacts do we have?" If you don't know what the problems are, you can't fix them!

There are various tools and techniques that organisations can use to identify and characterise the environmental issues associated with their activities. This needs to be done systematically to ensure that:

- nothing significant is omitted; and
- managing the most important issues is given priority.

Aims and Objectives of Impact Assessment

Environmental impact assessments may be used to identify environmental impacts from any activity or site, regardless of their size or nature. The objectives are to:

What environmental impacts do we have?

- Identify beneficial and negative environmental impacts of an activity.
- Suggest mitigation or control measures to prevent or reduce negative impacts.
- Identify appropriate strategies to monitor impacts and provide an early warning of any adverse changes.
- Incorporate environmental information into the decision-making process relating to development projects.
- Aid selection of the best option if alternatives are available.

EU Environmental Impact Assessment Directive (2011/92/EU)

It should be noted that the term Environmental Impact Assessment (EIA) has a special meaning when used in the context of the **EU EIA Directive**. This Directive requires formal, documented impact assessments of development projects likely to have significant environmental effects, prior to them being granted planning consent by local and/or national authorities. An environmental impact assessment must be carried out on the environmental effects of proposed major industrial or civil engineering developments, as specified in Annex 1 of the Directive - including, for example: crude-oil refineries, major power stations, iron and steel plants, chemical plants and major roads, railway lines and ports.

The **EIA Directive** applies to the UK, for example it is covered in England by the **Town and Country Planning (Environmental Impact Assessment) Regulations 2017**.

EIAs that meet the requirements of the Directive for planning purposes must follow a defined process:

3.1 Reasons for Carrying Out Environmental Aspect and Impact Assessments

Process and stages of environmental impact assessment

1. Screening	This stage is a formal decision about whether the development requires an EIA or not.
2. Scoping	Decide which environmental impacts are to be considered in detail. This will depend on the type of development, for example: • Air emissions? • Noise emissions? • Discharges to water?
3. Baseline Studies	For the impacts selected above: determine the current status. For example, what is the current air or water quality like?
4. Impact Assessment Significance	Consider whether your proposed or existing development will have a significant impact. For example, will your development cause the air quality to deteriorate significantly?
5. Mitigation	If there will be a significant impact, what can you do to reduce it? For example, can you fit abatement technology to control air emissions, or not operate at night to prevent noise nuisance to residents?
6. Application and Environmental Statement	Development of an environmental statement (different terminology may be used, but this is essentially a report) to document the EIA.
7. Monitoring	All developments or existing industries will need to monitor their significant impacts to ensure mitigation measures remain effective.

Let's look now at how we characterise environmental issues, and some key tools for identifying impacts.

Meaning of Aspects and Impacts

The photograph shows a coal-fired power station chimney. We can see that the chimney is emitting smoke and fumes into the surrounding air. This is clearly one way in which the facility is interacting with the environment - we call this an environmental aspect.

The international EMS standard **ISO 14001** defines an **environmental aspect** as an:

> "element of an organisation's activities or products or services that interacts or can interact with the environment".

So, one environmental aspect of the power station's activities is the emission of exhaust materials from its chimneys into the atmosphere. But why does this matter? What is the environmental damage, or impact, that might result from this aspect?

There is a very useful model that is often used to identify and characterise environmental impacts in these types of situation. It has the following elements:

Example of an environmental aspect

- **Source**: Is there a source of contamination? This might be a toxic chemical, a physical substance such as dust or grit, or energy in the form of heat, noise, or light.
- **Pathway**: Is there a route by which the contaminant can reach a receptor? For example, could the contaminant reach a receptor through the atmosphere, or possibly via a drainage system?
- **Receptor**: Is there something that can be harmed or damaged by this contaminant? This might be wild animals or plants, humans, ecosystems (e.g. rivers, forests), or global systems, such as the climate.

Reasons for Carrying Out Environmental Aspect and Impact Assessments 3.1

If we can make the link:

Source ⟶ Pathway ⟶ Receptor

then we have an **impact**.

ISO 14001 defines an **environmental impact** as:

> "change to the environment, whether adverse or beneficial, wholly or partially resulting from an organisation's environmental aspects".

If we apply the source-pathway-receptor model to the emission of exhaust materials from the coal-fired boiler in the power station, we find that this single aspect can have a number of quite different impacts. This is because there are a number of different contaminants associated with this aspect that can move via a number of different pathways to reach a number of different receptors.

For example, local residents could be affected by smoke and grit in the boiler exhaust, which passes up the chimney and into the local atmosphere and is then deposited over the neighbouring land. The smoke and grit could be a nuisance, e.g. creating dirty deposits on washing and cars, or people might suffer health effects if they are susceptible to breathing problems, e.g. due to asthma or bronchitis.

But there are also other types of contaminant (besides smoke and grit) contained in the exhaust emission. One of the main components of the power-station exhaust is carbon dioxide. This is one of the main greenhouse gases and contributes to global warming and climate change.

Another important component of the exhaust is the gas sulphur dioxide. If this gets into the atmosphere it can be transported many hundreds of miles by the prevailing wind. Sulphur dioxide can react with moisture in the atmosphere to create 'acid rain'. This has been shown to be damaging forests and lakes in countries as far away as Norway and Sweden.

So, the emission of exhaust materials from the power station's chimneys into the atmosphere - a single **environmental aspect** - can potentially result in a number of different environmental impacts, which involve different contaminants, pathways and receptors. In summary:

Source (Contaminant)	Pathway	Receptor	Environmental Impact
Smoke and grit particles	Released via chimney into the local atmosphere. Deposited on the ground close to the power station.	Local residents	Dirty washing
Smoke and grit particles	Released via chimney into the local atmosphere. Deposited on the ground close to the power station.	Local residents	Respiratory problems
Sulphur dioxide	Released via chimney into the atmosphere. *eg acidification of water course.* Carried by the prevailing wind and deposited as acid rain over Scandinavia.	Norwegian salmon	Declining fish stocks
Sulphur dioxide	Released via chimney into the atmosphere. Carried by the prevailing wind and deposited as acid rain over Scandinavia.	Norwegian trees	Forest die-back
Carbon dioxide	Released via the chimney into the global atmosphere.	Global atmosphere	Climate change

→ extent / location eg near / far.

3.1 Reasons for Carrying Out Environmental Aspect and Impact Assessments

Life-Cycle Analysis (Cradle-to-Grave Concept)

The source-pathway-receptor model that we looked at above is very useful in situations where an organisation is directly releasing contaminants or pollutants into the environment from its own activities, such as:

- exhaust emissions from a boiler; or
- discharges of liquid effluent from a manufacturing process.

Many organisations, especially in the service sector, may not, however, operate equipment or processes that make significant releases of contaminants directly into the environment. The environmental impacts of these organisations are more likely to be associated with the products and services that they supply or purchase.

Life-Cycle Analysis (LCA) is a tool to identify and measure the environmental impact of a product or service throughout its life cycle - from cradle to grave. This information can then be used to inform the decision-making process regarding new products or to make changes to materials used in existing products.

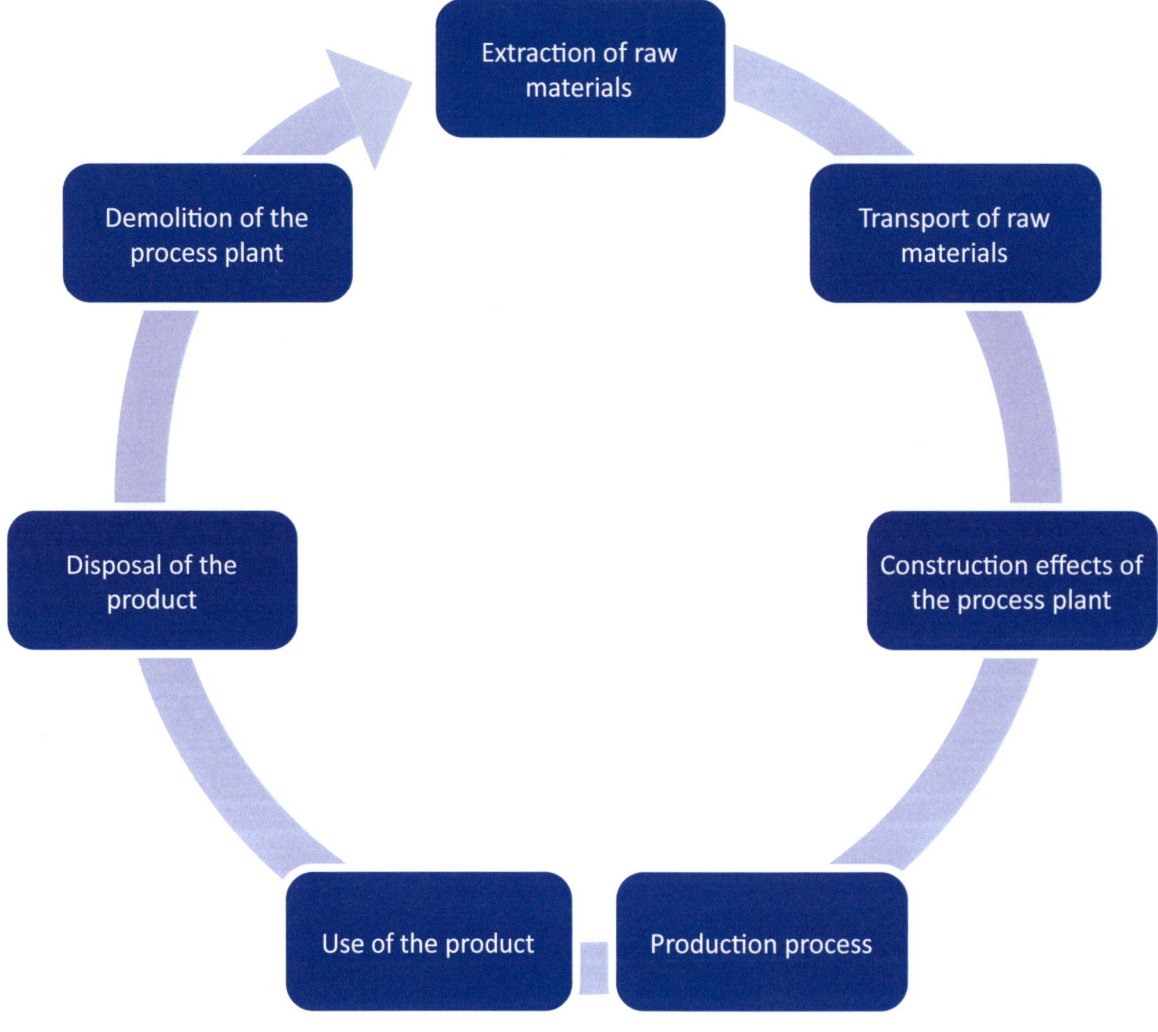

Possible life cycle of a product

The **ISO 14040** series provides guidance on aspects of LCA. The idea is to collect data on inputs at each stage of the life cycle, such as the use of raw materials and energy, and outputs for each stage, such as emissions to air, water and waste products. This allows an evaluation of the environmental impact of each stage in the life cycle, as well as for the product as a whole.

Reasons for Carrying Out Environmental Aspect and Impact Assessments — 3.1

This cradle-to-grave approach helps identify the total impact a product has on the environment. One example of this is the often-discussed argument over the use of disposable or washable nappies for babies. For some years now, there has been a belief that washable nappies are significantly better for the environment than disposable nappies. This is because disposables usually end up in landfill and can take hundreds of years to decompose, thereby continuing to take up space and contribute to the generation of landfill gas containing methane – a known contributor to climate change. However, the counter argument to this is that washable nappies consume large amounts of energy during their lifetime because of the need to wash them frequently and often at high temperature. In practice it is likely that there is no significant difference between the options over a full life cycle.

Two further examples of the use of LCA are the comparison of the environmental impacts of low ethanol-content diesel blends (e-diesel) with traditional fuels, and what type of packaging materials have the least impact overall on the environment.

- Increased efficiency and saving - the process allows the tracking of energy flows through a production process and the highlighting of areas where savings can be made.
- Product marketing - information gained during the process can be used to highlight the environmental benefits of the product, thereby attaching a potentially new Unique Selling Point (USP) to the product and differentiating it from other, similar products in the marketplace.

MORE...

For information on 'green' claims, go to:

www.gov.uk/government/publications/make-a-green-claim

Principles and Techniques of LCA

The LCA process can be applied to a product whenever there is an input to or output from the product. A full LCA of even a simple product is a long and complex process and it is beyond the scope of this course to discuss it in detail.

TOPIC FOCUS

ISO 14040 identifies four stages in the **LCA process**:

- **Definition of goal and scope** - an important and early decision is the purpose and coverage of the LCA. Which stages of the life cycle and which inputs and outputs are to be assessed?
- **Inventory analysis** - this is the data collection stage, where inputs and outputs are identified and quantified.
- **Impact assessment** - once the data has been collected and quantified in the inventory analysis stage, there will need to be some assessment of the impact of those inputs and outputs on the environment. This may involve attaching the inventory data to a known impact category. For instance:
    - Oxides of sulphur and nitrogen may be allocated to an impact such as acid rain.
    - Greenhouse-gas emissions may be allocated to a global warming category.
- **Interpretation of the results** - this final stage brings all the data gathered in the previous stages together and allows for conclusions and recommendations to be drawn from the results. The outcome of this process may be changes to the product design, or how it is made. It may also provide marketing material to help position the product more favourably in the marketplace.

STUDY QUESTIONS

1. Outline what is meant by the terms 'environmental aspect' and 'environmental impact'.
2. What is life-cycle analysis?

(Suggested Answers are at the end.)

3.2 Types of Environmental Impact

Types of Environmental Impact

IN THIS SECTION...

Environmental impacts can be categorised in a number of ways:

- whether they are direct or indirect;
- by the media they pollute; or
- whether they are positive or negative impacts.

Direct and Indirect Impacts

Impacts can arise as a direct result of an organisation's activities. For example, in the case of the coal-fired power station that we looked at earlier, the operator of the power station clearly has management control and responsibility for the facility. The environmental impacts associated with the power station are therefore the operator's **direct** environmental impacts.

The more electricity we use, the more emissions the power station generates

 But we all use electricity, don't we? We create a demand for electricity, and the more electricity we use, the more emissions the power station generates. We also have some capacity to reduce the environmental impacts associated with electricity generation by using less energy. We therefore accept some responsibility by recognising the consumption of electricity as an **indirect** environmental impact of our activities.

Indirect environmental impacts are associated with many of the goods and services that we buy. For example:

- Use of a third-party distribution contractor (air pollution from trucks).
- Purchase of paper (cutting down forests; pollution from paper mills).

For many organisations, especially in the service sector, indirect environmental impacts may therefore have a very high significance in their environmental programme.

Cumulative Impacts

Such impacts are those that occur from incremental changes over time. Individual impacts may be relatively minor but in combination with others they could have a significant environmental impact. Examples include loss of soil quality, climate change and damage to habitats.

Positive and Negative Effects

Air and Atmosphere

This is almost impossible to control once a pollutant is emitted to the atmosphere. The effects may not only be local but can be national or international (transboundary). Air pollution can also have significant adverse health effects.

Land

This can lead to restricted use of land in the future or entail significant clean-up costs. Pollution of land can also lead to pollution of watercourses, either by percolation through to groundwater, or by surface water run-off to rivers and streams.

*if hire others becomes an indirect impact eg own van = direct, hire in van = indirect.

Water

Once in an aquatic environment, pollution can travel long distances and cause adverse effects significant distances from the original source. If groundwater becomes polluted, it may be extremely difficult to effectively clean the water so that it is wholesome again. Oceans may be polluted by various materials such as plastics, heavy metal and nutrients.

Human Communities

Local residents may benefit from development. For example, a mixed development that combines commercial, cultural, institutional, or industrial uses can provide a number of benefits for a neighbouring residential community, such as:

- Creation of jobs.
- Reduced travel between homes, workplaces and retail and leisure facilities.
- Provision of more open spaces and pedestrian and cycle paths.
- Improved landscaping through tree planting and the creation of nature habitats.

However, there may also be negative impacts on the local community:

- Noise and vibration.
- Increased traffic causing congestion, dust and odour.
- Loss of open space.
- Polluted watercourses.
- Loss of visual amenity - a wide-open view of the countryside will be preferable to a large industrial development.

Polluted sea shore, as well as loss of visual amenity

Effects on the Ecosystem and Species

Ecosystems are a collection of individual habitats (places inhabited by various forms of wildlife). Damage to them can have a significant effect on biodiversity and may impact on the services that such ecosystems supply (such as flood prevention).

On the positive side, however, certain developments, such as the remediation and development of a contaminated site (e.g. an old landfill), or planting trees on previously arable land, might have several benefits in terms of the restoration or creation of wildlife habitats.

Afforestation may also have several benefits beyond habitat creation, such as:

- Acting as a carbon sink – carbon dioxide sequestration. (Trees use carbon to create energy and naturally remove carbon dioxide from the atmosphere for this purpose. The tree acts as a long-term carbon-storage reservoir.)
- Improved water quality (remember, arable land may be being displaced, reducing the amount of agricultural fertiliser ending up in local rivers).
- Reduced soil erosion (rain infiltration rates tend to be lower, so there is less surface run-off).
- Flood control (again, because surface run-off is reduced).

All of these can affect both humans and animals.

3.2 | Types of Environmental Impact

STUDY QUESTION

3. Consider a vehicle repair garage. Which of the following environmental aspects will have impacts that are direct, and which will have impacts that are indirect?

    - Air emissions from running vehicle engines in the workshop.
    - Air emissions from a waste oil burner used to provide heat for the workshop.
    - Methane gas generated from a landfill site that disposes of the garage waste.
    - Accidental spillage of oil into the public sewer at the garage.
    - Use of electricity from the energy provider to heat offices.
    - Purchasing of office desks, made from non-renewable hardwoods.
    - Contamination of the land through spillages of oil and fuel in the workshop.
    - Water discharges caused in the manufacture of engine parts.

(Suggested Answer is at the end.)

Nature and Key Sources of Environmental Information

IN THIS SECTION...

- Various internal and external sources of environmental information are available to assist an organisation in assessing impacts.
- Supply-chain issues should also be considered.

Internal to the Organisation

To identify significant aspects and impacts, you need information about your organisation's activities.

> **TOPIC FOCUS**
>
> *[handwritten note: reactive data from incidents.]*
>
> **Sources of Environmental Information Internal to the Organisation**
>
> - **Inspection/audit reports**: identify any evidence of defects and a lack of suitable controls.
> - **Incident data and investigation reports**: indicate where there has been a failure in a process or procedure. The frequency and extent of such a failure will help determine the probability and severity of such an incident. Investigation will help determine causes and additional controls.
> - **Maintenance records**: provide information on machine reliability and types of failures experienced in the past. This can be used as an indication of what may occur in the future and therefore what precautions need to be put in place.
> - **Job/task analysis**: although usually used to identify safety hazards, a thorough analysis of tasks - by breaking them down into smaller steps - may help identify potential environmental impacts from the activity.
> - **Environmental monitoring data**: the results can show trends, such as pollution levels increasing slowly over a period of time, and identify any times of increased risk when it may be necessary to implement stricter controls.
> - **Raw-material usage and supply**: volumes of raw material used should be consistent with the volumes of final product produced. If there are significant changes, this may be an indication of a problem in the production process. Consideration should also be given to the sources of supply and the potential impacts this may have, e.g. using timber from a sustainable source.
> - **Environmental permits**: will provide detailed information on what activities may be carried out and the levels of pollution that are permitted from those processes. They will usually also include details on how frequently monitoring must take place and what parameters must be monitored, such as total volume, rate of discharge, suspended solids, etc.

3.3 Nature and Key Sources of Environmental Information

External to the Organisation

> **TOPIC FOCUS**
>
> **Sources of Environmental Information External to the Organisation**
>
> - **Manufacturers' data**, including information such as Material Safety Data Sheets and operating or maintenance instructions.
> - In many countries, **legislation** is made freely available online, or available to purchase as a hard copy.
> - **Enforcement bodies** publish guidance documents on compliance with environmental law and promoting good practice.
> - **Government-supported organisations** whose role is to support and encourage environmental improvements in specific areas. Examples of these in the UK include the Waste and Resources Action Programme (WRAP) (www.wrap.org.uk) and the National Industrial Symbiosis Programme (www.nispnetwork.com).
>
>   The European Environment Agency body produces information for member states on environmental issues.
>
> - **Trade associations** (such as the Chartered Institution of Wastes Management (CIWM), the Royal Institution of Chartered Surveyors (RICS), the Mineral Products Association and many others in the UK, for example) can provide specific advice and information on the areas of expertise in which they operate.
> - **Professional institutions** (such as the Chartered Institution of Water and Environmental Management (CIWEM) and the Institute of Environmental Management and Assessment (IEMA) in the UK) exist to provide support to professional members and to promote a higher standard of training and competency for those working in these areas. Many of them also have a wide range of consultancy services and technical information available.
> - International Organisation for Standardisation/recognised European standards organisations (e.g. European Committee for Standardisation)/British Standards Institution publish the standards such as ISO 14001:2015 for environmental management systems and often have guidance documents to support these standards.
> -  **Commercial organisations** such as Barbour, Technical Indexes, etc. all offer either online, CD or book-based systems for accessing legislation and guidance. Some of them also offer specific helpline services and documents, such as checklists and form templates.
>
>   *eg Croner*
>
> - **Encyclopaedias and textbooks** are also available.

STUDY QUESTION

4. List three sources of environmental information that are internal to the organisation and three sources of information that are external to the organisation.

(Suggested Answer is at the end.)

Identifying Environmental Aspects and Associated Impacts

IN THIS SECTION...

- Impact assessment can be linked to the initial environmental review undertaken in **ISO 14001**.
- Different operating or unplanned conditions should be considered when assessing environmental impact.
- The concept of Source, Pathway, Receptor is critical in assessing environmental impact.
- Semi-quantitative assessment is a common method for determining the significance of an environmental impact.
- Environmental impact assessments should be recorded and reviewed periodically.

Implementing an EMS

When implementing a new Environmental Management System (EMS) from scratch, an organisation's significant aspects and impacts would normally be identified at the initial environmental review stage (see Element 2). The policy and objectives would then be developed realistically based on knowledge of the specific environmental risks arising from the organisation's activities.

Identifying Environmental Aspects

The impact types discussed earlier are usually considered within a number of contexts:

- **Direct/indirect**: as discussed earlier.
- **Normal/abnormal conditions**: all planned activities need to be considered. These will include not only those associated with normal running but also those associated with the non-routine ('abnormal'), such as maintenance and cleaning.
- **Accidents/incidents/emergencies**: reasonably foreseeable incidents should also be considered, e.g. fire or chemical/oil spillage.
- **Past/future activities**: you should consider the impact from past and planned activities. For example, past land contamination has impacts that continue into the present. Business plans, such as increasing production, will also have future impacts that should be taken into account.

Fire can be a reasonably foreseeable incident

Control and Influence

ISO 14001 states that identification of environmental aspects should be carried out for activities, products or services that can be controlled by an organisation, or the activities, products or services that an organisation is expected to have reasonable influence over. It allows identification of aspects that have a significant environmental impact.

Organisations therefore have some flexibility when identifying the scope of the EMS:

- Control - these issues are likely to result from direct activities associated with emissions from processes and compliance issues.
- Influence - this is much broader and mainly includes indirect issues, such as suppliers, contractors and customer activities. Influencing can be achieved by communication and supply-chain pressure.

Many organisations will initially concentrate their effort on identifying direct activities that they can control. As an EMS matures, an organisation will consider more indirect activities. For banks, insurance companies, etc. the significant aspects are mainly likely to be surrounding indirect activities.

3.4 Identifying Environmental Aspects and Associated Impacts

Life-Cycle Perspective

ISO 14001 states that a life-cycle perspective should be considered when determining environmental aspects and impacts. This means identification of aspects and impacts from the cradle to the grave (from the development of raw materials all the way to the final disposal of the product) and is likely to cover impacts associated with suppliers, distribution and waste disposal.

Determining Associated Environmental Impacts

Concept of Source, Pathway, Receptor when Assessing Environmental Risk

The concept of Source Pathway Receptor (SPR) - sometimes referred to as Source Pathway Target (SPT) - that we introduced earlier is fundamentally useful in assessing environmental risk. The SPR approach can be used to identify potential effects on any environmental media (air, water, and land).

> **TOPIC FOCUS**
>
> **Example of the SPR Approach: A Petrol-Filling Station**
>
> There are numerous sources, pathways and receptors here, of which the following are examples:
>
> SPR analysis
>
> | Source | Pathway | Receptor |
> |---|---|---|
> | Underground fuel tank | Product loss and dissolution in groundwater | Groundwater in aquifer |
> | | Vapour transport through soil | Humans |
> | Fuel dispenser | Air-inhalation | Humans |
> | Spills by users | Forecourt drains | Local watercourses |

Identifying Environmental Aspects and Associated Impacts | 3.4

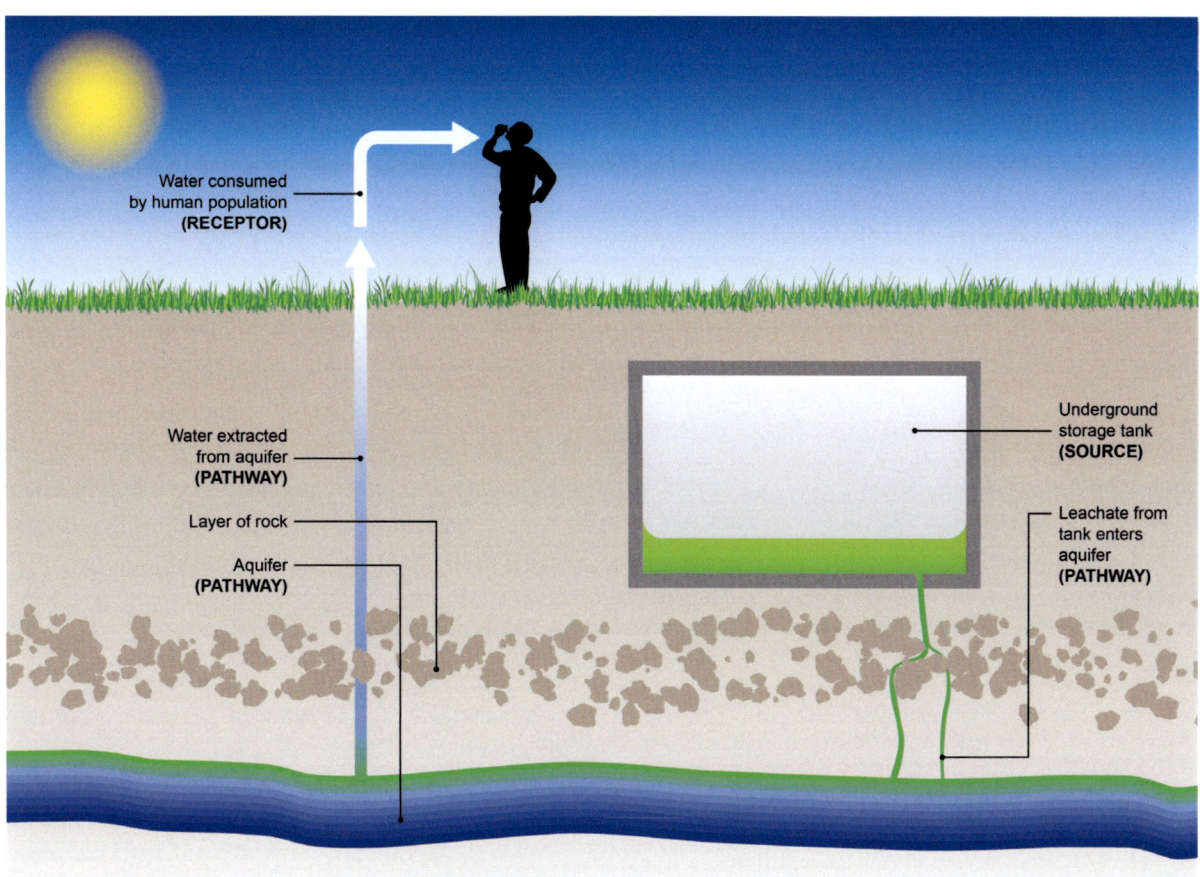

Source, pathway and receptor - an example

Note: Pathways may also be receptors (such as in the case of a watercourse), and receptors can include people or land of varying sensitivities.

You will appreciate that a great amount of analysis is required to understand environmental harm. Although the receptor may be a watercourse, the effect may be on the fish or invertebrate life in that watercourse, or to the humans or others who have use of that water, whether as drinking water for people or animals, or for recreational or industrial use.

Normal/Abnormal/Emergency Conditions

All planned activities need to be considered. These will include not only those associated with normal running but also those associated with the non-routine ('abnormal'), such as maintenance and cleaning. Reasonably foreseeable emergency situations should also be considered, e.g. fire or chemical/oil spillage.

Other Considerations

As we considered earlier, it is important that the following are taken into account when determining impacts:

- Positive and negative impacts.
- Direct, indirect and cumulative impacts.

The evaluation should not just include present activities but also past activities (e.g. spillages, land contamination), and future activities (e.g. new developments or business plans to increase production). The adequacy of current controls will also affect the likelihood of an impact occurring and as such should form part of the evaluation of whether an impact is significant.

3.4 Identifying Environmental Aspects and Associated Impacts

Specific Impacts

International Impacts

As we discussed earlier, there are numerous international impacts that should be considered when carrying out an environmental impact assessment. These include:

- Climate change.
- Ozone depletion.
- Acid deposition.
- Water pollution (where watercourses pass international boundaries).
- Waste disposal (where waste is transported from one country to another), etc.

Resource Abstraction

Taking resources from the environment can have numerous impacts on the environment. Deforestation - the removal of naturally-occurring forests by human activities such as logging or the burning of trees - is associated with significant environmental problems. Deforestation may occur for a number of reasons, such as clearing land for cattle, settlements or agricultural plantations and the use of wood for charcoal.

The removal of trees without sufficient replanting leads to several problems, including:

- Damage to habitats.
- Biodiversity losses.
- Soil erosion, which allows fertile soil to be washed into rivers, leaving behind wastelands.
- Removal of a key carbon sink, increasing the amount of carbon dioxide in the air.

Pollution from Mining

The impacts on the environment of metal extraction, for example, can be significant and include:

- Deforestation to make way for mines.
- Rock waste from surface mining which is typically deposited on land close to the mine, covering areas that are vegetated.
- Mine waste (known as tailings); this can leach into nearby rivers and other types of watercourses, causing them to become clogged and flooded.
- Metals present in the mine waste which can pollute rivers and other surface waters, seriously affecting aquatic life. Sulphur in mines can combine with water to form sulphuric acid, creating acid mine drainage.

Transport

In the control of air pollution, transport effects include the emission of combustion gases, such as carbon dioxide, carbon monoxide, nitrogen and sulphur oxides (NO_x and SO_x), particles and, in lesser quantities, up to 40 other gases, such as butadiene and benzene.

Other effects of traffic should not be overlooked, including:

- Nuisance caused by noise or dust.
- Congestion causing air pollution and nuisance.
- Changes to the landscape affecting aesthetics.
- Land-take causing reduction in diversity.
- The effects of refuelling causing water pollution from spillage.

Waste Disposal

The impacts of waste disposal can be significant and we will cover these in more detail later. They include:

- Noise from waste transportation and site activities causing nuisance.
- Odours from landfill sites or waste incineration causing nuisance.
- Dust and litter causing nuisance.
- Release of methane containing landfill gas causing climate change and presenting a fire and explosion risk to those near the site.
- Leachate discharged from a landfill causing water pollution.

Identifying Receptors at Risk

It is essential that potential receptors are identified early in the process. Some of these are discussed below.

Flora

Carry out an ecological assessment on the surrounding flora such as reedbeds, hedgerows and local woodland. This is likely to require specialist knowledge and monitoring to be conducted over at least a 12-month period to account for seasonality.

Fauna

Carry out an ecological assessment on the surrounding fauna, including habitat of bats, birds and badgers. This should also include a protected species survey that would need to be completed at the appropriate time of year for invertebrates, small mammals (including bats, dormice, water voles) and nesting birds.

Carry out ecological assessments on the surroundings

Watercourse

Assess the potential impacts on groundwater and surface drainage ditches and conduct a survey of water abstractions to establish the number of licensed and unlicensed abstractions and their proximity to the site.

Local Populace

Impacts could include air and climate, noise and vibration, cultural heritage, landscape and visual impacts, as discussed earlier. The impacts on indigenous people must be considered, where relevant, when assessing the risk to the local populace.

Identification of Aspects and Impacts

The organisation's environmental aspects and associated impacts now need to be identified, making use of the SPR model for direct aspects/impacts, and the Cradle-to-Grave approach for indirect aspects/impacts, as appropriate.

3.4 Identifying Environmental Aspects and Associated Impacts

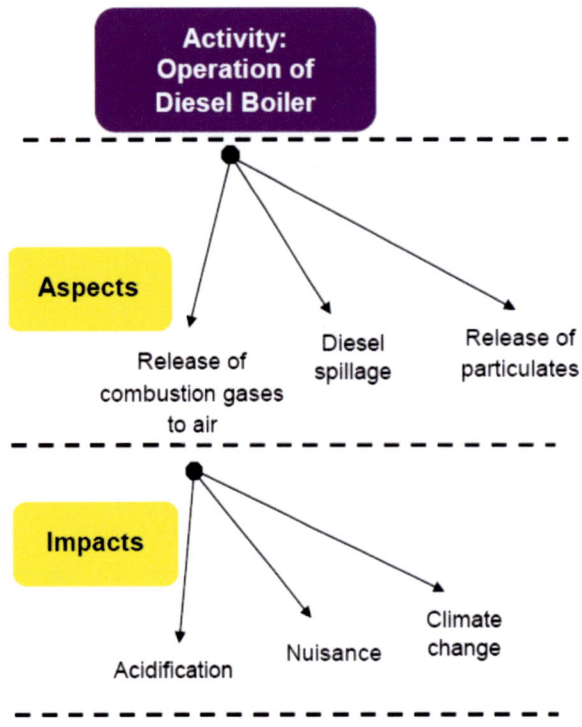

Example of aspects and impacts for the operation of a diesel boiler

There are many ways to identify aspects and impacts. A common approach is to break a site down into functions or processes and then look at the activities or services associated with each. For each activity, you can then consider its inputs (raw materials, etc.) and outputs (solid waste, etc.). A checklist can be used for this purpose.

Once these have been identified, it is necessary to establish which of those aspects are significant and need to be controlled. This is discussed below.

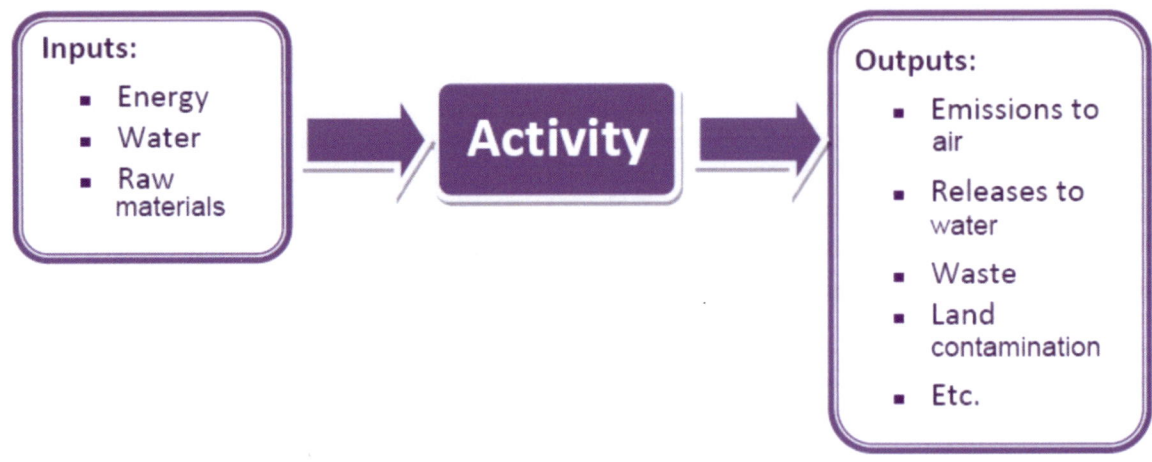

Input/output method of identifying environmental aspects

Evaluating Impact and Adequacy of Current Controls

> **DEFINITIONS**
>
> **QUANTITATIVE**
>
> Usually measured in some way and quantified, e.g. measured emissions from a process.
>
> **QUALITATIVE**
>
> Based on some quality rather than quantity - usually a subjective judgment, e.g. 'good' and 'bad', or 'low', 'medium' and 'high'.

Having systematically identified all of the environmental impacts of all of their activities, most organisations will be faced with a very long list. It will usually not be feasible to address all of these impacts immediately, and some impacts may be so small as to be considered trivial.

Clearly, some process is needed to analyse the long list of impacts and to determine which of these are significant and should therefore be given priority.

There is no single system for determining the significance of impacts. But many organisations already have a well-established system for health and safety risk assessments, which can easily be adapted to cover environmental impacts.

When evaluating the significance of environmental aspects and impacts the following factors should be considered:

- Scale and severity of the impact.
- Duration of the impact.
- Business concerns.
- Sensitivity of the receiving environment.
- Legal and contractual requirements.
- Needs and expectations of interested parties.
- Effect on public image.

The most common methods for determining significance are semi-quantitative assessments, which use a scoring matrix to combine relative scores for the likelihood that an impact will occur and its severity if it does so.

This approach generally works well, but it must be emphasised that a good deal of judgment is required, and although numbers are used, these are only indicative and do not confer absolute validity.

An example of a matrix system is shown below. The way it works is that each impact is given two scores on a scale of 1 to 5. The first score expresses the relative likelihood that an impact will occur, and the second, the severity of the impact should it actually happen. Simple criteria have been documented to guide the award of scores.

The individual scores for likelihood and severity are then multiplied to give an overall assessment score. The overall scores can then be filtered into 'High', 'Medium' and 'Low' bands to determine the level of significance.

3.4 Identifying Environmental Aspects and Associated Impacts

Assessment Scores	
Severity	**Likelihood**
No impact - 1	Rare - 1
Minor and temporary impact - 2	Unlikely - 2
Minor but permanent impact - 3	Probable - 3
Major impact - 4	Very likely - 4
Major impact in breach of legislation with risk to health - 5	Certainty - 5

Significance Factor	
Severity × Likelihood then gives overall Significance Factor (minimum = 1; maximum = 25)	

Significance Assessment	
Low	1 to 7
Medium	8 to 15
High	16 to 25

Semi-quantitative environmental risk assessment matrix

The risk assessment will, of course, look at existing controls and, if considered inadequate in relation to the risk, recommend further controls.

Risk and Opportunities

It is important to understand the opportunities that may occur when controlling or reducing significant environmental impacts. Primarily significant impacts will be identified, following which the opportunities that occur for managing those impacts can be identified. These include:

- Financial savings - through reductions in waste, energy and raw materials use.
- Regulatory - having good standards of environmental management will reduce the probability of an organisation breaching legal requirements and the consequent associated adverse issues such as prosecution, regulatory cost and negative corporate image.
- Sales improvement - where an organisation is considered to have a high level of environmental performance it can lead to an increase in sales and/or profit through access to new markets and increased customer share.
- Investment - investors are more likely to lend at a preferential rate to organisations that have a lower overall risk. Environmental risk is an important form of risk that can have a financial bearing on an organisation.
- Staff - potential employees are more likely to want to work for an organisation that has reputable values.

Recording Significant Aspects and Impacts

Environmental aspects and impacts are often recorded in the form of a register. Although there is no set format for an aspects and impacts register, as a minimum they will often contain the following information:

- Activity that has been assessed.
- Environmental aspects associated with the activity.
- Environmental impact associated with each aspect.
- Identification of whether the aspect is from normal, abnormal, or emergency context.
- Identification of the results of the assessment against significance criteria, e.g. likelihood, consequence and total scores if using a semi-quantitative method to assess significance.

Reviewing

Any risk assessment is, of necessity, a snapshot of the situation as it is at the time the assessment takes place. It may be necessary to review the assessment if:

- An incident relevant to the assessment occurs, particularly if there appear to be inadequate controls.
- Processes and/or equipment changes.
- Staff changes, particularly if there is a loss of experience and/or historical knowledge.
- Legislation changes, imposing more onerous or new requirements.
- There has been a lapse of time since the last review - technology moves on, and what is considered the best available technique, or best practicable environmental option today, may not be in the future.

> **STUDY QUESTION**
>
> 5. Identify three criteria that might be used in deciding whether an environmental impact is significant or not.
>
> (Suggested Answer is at the end.)

Summary

This element has dealt with the principles and practice of Environmental Impact Assessment (EIA).

In particular, this element has:

- Shown that the main objectives of an EIA are to identify environmental impacts of an activity, suggest measures to prevent or reduce negative impacts, identify monitoring strategies, and incorporate environmental information into the decision-making process.
- Outlined the stages of an EIA, including screening, scoping, baseline studies, impact assessment significance, mitigation, application and environmental statement, and monitoring.
- Explained how Life-Cycle Analysis identifies and measures the environmental impact of a product or service throughout its life cycle - from cradle to grave. The information can be used to inform the decision-making process regarding new products, or to make changes to materials used in existing products.
- Shown that environmental impacts can be categorised in the following way:
    - Direct and indirect.
    - Contamination of the atmosphere.
    - Contamination of land.
    - Contamination of the aquatic environment.
    - Positive and negative effects on the community and on the ecosystem.
- Identified many sources of information available relating to an organisation's activities:
    - Internal, e.g. inspection/audit reports; maintenance records, etc.
    - External, e.g. manufacturers' data; enforcement bodies, guidance documents, etc.
- Described the principles and practice of impact assessment, including:
    - The concept of Source, Pathway and Receptor - identifying potential effects on any environmental media.
    - Examples of specific impacts (including international impacts, resource abstraction, pollution from mining, transport and waste disposal).
    - Identification of receptors at risk (flora, fauna, etc.).
    - Identification of the organisation's environmental aspects and impacts and significant environmental impacts (impacts may not always be negative).
    - Evaluation of the significance of each impact (e.g. severity × likelihood).
    - Considering the activities of suppliers, in particular the significant indirect environmental impacts for retailers and other service sector organisations.
    - Recording any significant findings.
    - Reviewing the assessment as and when necessary.

Exam Skills

Question

> **Scenario**
>
> You have been asked to implement an environmental management system for a large supermarket chain. The chain operates hundreds of supermarkets in different locations in a developed country.
>
> **Task**
>
> What measures could be implemented to reduce the environmental impact of the chain's stores?
>
> **(4 marks)**

Approaching the Question

Think about the steps you would take to answer the question:

- Read the scenario carefully. With this question you need to develop measures that could be implemented 2o reduce the environmental impact of the chain's stores. Note: the question states "stores" only.
- Now look at the task – prepare notes (and examples) on:
  - Supermarkets' aspects and impacts (the key risks need to be determined to provide a systematic way of identifying improvement measures).
  - Measures to reduce impact.
- Consider the marks available. In this case, there are 12 marks available so you should provide at least 12 measures that could be implemented. However, these should be split fairly evenly across a number of different impact groups.
- Read the scenario and task again to make sure you understand them and have a clear understanding of the impacts of supermarket stores and the ways to reduce their impact.
- Jot down an outline plan - this might include:
  - Energy use - climate change, acid rain, renewable sources, lighting improvements, on-site generation, insulation.
  - Emissions to atmosphere from vehicle exhausts - climate change, acid rain, alternative fuels, regular maintenance, driver training, car sharing.
  - Waste - impacts of landfill, waste minimisation, re-use of wastes, long-lasting bags, segregation, recycling.
  - Packaging - resource consumption, climate change, landfill, pressuring suppliers, easily recycled, information on recyclability, sustainable sourcing.

Now have a go at the question yourself.

Exam Skills

Example of How the Question Could be Answered

Impacts associated with energy use can include climate change and acid-rain production. These can be reduced by the store purchasing electricity that has been generated from a renewable source, such as hydropower, solar power, etc. The organisation could also improve its lighting by fitting more efficient lighting (e.g. compact fluorescent bulbs), optimising the amount of lighting and fitting lighting controls in appropriate areas such as movement sensors in corridors. The organisation could also implement on-site alternative energy-generation systems, such as using solar photovoltaic panels on store roofs that will generate electricity for the site with surplus being passed to the national grid. The organisation could also fit insulation to its stores, as this will mean that less heating of the building will be required during colder periods of the year.

Emissions from vehicle exhausts can have numerous impacts, such as climate change, acid rain, reduction in biodiversity, etc. By using alternative fuel vehicles for deliveries or supply, this will reduce the amount of carbon dioxide emitted. Electric vehicles, for example, emit lower levels of greenhouse gases when combusted than petrol or diesel. The organisation could also ensure that regular maintenance is undertaken on company vehicles. Simple measures, such as regular tyre-pressure checks, can result in much improved efficiency. Driver training could also be provided in techniques to reduce fuel use and emissions, such as avoiding inappropriate acceleration. The organisation could also promote the use of car sharing to staff by operating a car-share system.

Numerous wastes can be produced by the organisation. It is likely these will end up in landfill causing numerous impacts, such as nuisance, climate change and water pollution. Implementing a waste minimisation programme would assist in reducing the amount of waste that goes to landfill. This might include the organisation re-using certain items of packaging for other uses at the site, e.g. boxes could be provided for customers to carry goods. Offering more long-lasting carrier bags and giving customers an incentive to re-use carrier bags could also be implemented (e.g. points on a loyalty card). The organisation could also segregate its general waste, such that plastics, cardboard, etc. are separated and sent for recycling. The company could also pressure suppliers to reduce the packaging on products that are deemed to be over-packaged.

Impacts associated with packaging are many and can include resource consumption, climate change, impacts of landfill, etc. Ways to reduce packaging include pressuring suppliers of products to optimise packaging where it is not required. The company should ensure that the packaging chosen can be easily recycled. The packaging should also contain information (such as text and/or symbol) that identifies whether it can be recycled. The store should also ensure that packaging is made from a sustainable source, e.g. in the UK Forest Stewardship Council (FSC)-accredited cardboard should be used.

Element 4

Planning for and Dealing with Environmental Emergencies

Learning Outcomes
- Support environmental emergency planning.

Learning Objectives

Once you've read this element, you'll be able to:

1. Explain the importance of environmental emergency planning.

2. Describe suitable emergency preparation and responses.

Contents

The Importance of Environmental Emergency Planning	**4-3**
The Effects of Unplanned Incidents on the Environment	4-3
General Duty or Responsibility Not to Pollute	4-3
Requirement of the Environmental Management System	4-3
Need for Prompt Action to Protect People, the Environment and Organisational Assets	4-4
Immediate Risks	4-4
Long-Term Risks	4-4
Emergency Preparedness and Response	**4-5**
Recognising Risk Situations and Action to Take	4-5
Emergency Response Plans	4-7
Information and Training for Internal and External Interested Parties	4-9
Review and Continual Improvement of Emergency Response Plans	4-12
Summary	**4-13**
Exam Skills	**4-14**

The Importance of Environmental Emergency Planning

IN THIS SECTION...

- There is a general duty or responsibility not to pollute the environment, so emergency plans should include procedures to control environmental hazards.
- A prompt response during an emergency can substantially reduce the environmental consequences of the incident.
- Strict liability is sometimes placed on organisations in relation to environmental damage. This means that the enforcing authority does not need to prove intent or negligence to successfully prosecute.
- Poorly managed environmental incidents usually require extensive and expensive clean-up and the reputation of an organisation can be seriously impacted.

The Effects of Unplanned Incidents on the Environment

Unplanned incidents can result in significant environmental impacts. For example, the Deepwater Horizon oil spill that took place in 2010 in the waters of the Gulf of Mexico was caused by an explosion on an oil exploration rig, and resulted in a massive oil spill that contaminated marine and coastal ecosystems for hundreds of kilometres. Thankfully, emergencies that result in major environmental impacts are rare, but minor emergency incidents, such as small spillages of fuel, are quite common and a significant source of local pollution.

The risk of a major, or minor, emergency causing knock-on environmental impacts can be greatly reduced by good planning and management.

Deepwater Horizon

General Duty or Responsibility Not to Pollute

An organisation has a general duty or responsibility not to pollute the environment, so even if the primary concern is the protection of people, an emergency plan should include the environmental hazards and procedures to control them.

All organisations should have emergency plans in place. The size, complexity and primary aim of these plans will vary depending on the size and nature of the organisation. Office-based companies may have relatively simple plans with a primary aim of protecting life and ensuring the business is able to continue following the emergency.

Requirement of the Environmental Management System

Any Environmental Management System (EMS) should identify the significant aspects and impacts that can arise from the organisation's operations, and this should include potential impacts under emergency conditions. **ISO 14001** also requires an organisation to:

- develop and maintain documented processes to identify and respond to accidents and emergencies; and
- prevent or reduce environmental impacts that are associated with them.

Identifying potential emergencies during the development of the EMS allows the procedures to be communicated, along with those for the normal and abnormal situations, and this helps to ensure that employees are fully aware of the requirements.

4.1 The Importance of Environmental Emergency Planning

Need for Prompt Action to Protect People, the Environment and Organisational Assets

During an emergency, there are usually a lot of different things happening at once and quickly. An emergency is a fluid situation; full information is often unknown at the start of the incident. Nevertheless, situations have to be recognised and evaluated quickly and decisions made to mitigate potential consequences to people, the environment and organisational assets. For example, drains may have to be blocked and valves closed to stop a spillage escaping from a site into controlled waters. The middle of an emergency is not the time to have to make critical decisions that have not been considered beforehand. There is usually very little time to start finding new information, such as to where certain drains run.

The emergency services will not arrive immediately. There is therefore a window where prompt action may significantly reduce the potential environmental damage.

Immediate Risks

Unplanned events such as emergency situations can have significant negative effects on business, making it very difficult or impossible to carry out normal activities. This can have some negative effects on organisations, including loss of business and potential site shutdown. Clean-up of a pollution incident could be costly and will significantly hinder the organisation's productivity. With good planning, however, steps can be taken to prevent an emergency from occurring and to reduce the impact of an emergency should it happen.

Long-Term Risks

Risks of Prosecution and Other Costs

Emergency situations are clear evidence that somewhere, something has gone wrong - either there has been equipment failure or someone has made a mistake. Regulatory regimes for environmental protection often impose a strict liability on organisations for the damage they cause to the environment. This means that it is not necessary for the regulating authority to prove any intent or negligence on the part of the organisation, but merely to show that they caused the damage.

- In some cases, fines can be unlimited. There is also the possibility of an individual being prosecuted and, if found guilty, facing a custodial sentence.
- Clean-up costs may also be awarded against the organisation that has caused the damage. In some cases, these may be considerably higher than the fine itself.
- Civil claims may be made by those who have suffered some loss as a result of the incident.

Reputational Issues

Many organisations value their reputation above almost anything else. It is what they trade on, it is their image in the marketplace and possibly the most valuable (yet most easily lost) asset an organisation can have. With the current focus on environmental issues, any organisation seen or believed to be causing damage to the environment is likely to lose favour with significant sections of the market, including potential investors. If the environmental incident takes place at a critical time for the organisation, such as when negotiating finance for expansion or a major contract to supply a new customer, adverse publicity caused by the incident may result in the finance or the contract being withdrawn and, together with the civil and criminal sanctions imposed, may potentially lead to the downfall of the organisation.

> **STUDY QUESTION**
>
> 1. Briefly explain three reasons why organisations should have emergency plans in place.
>
> (Suggested Answer is at the end.)

Emergency Preparedness and Response

IN THIS SECTION...

- Some organisations, such as large chemical works, are required to produce an emergency plan, which addresses environmental issues. These organisations are controlled under national laws made to comply with **Directive 12/18/EU** on the control of major accident hazards involving dangerous substances (known as the **COMAH Directive**).
- Other organisations may produce emergency plans in response to EMS requirements, planning law, or because it represents good practice.
- Emergency plans should consider: possible accident scenarios, the predicted environmental effects, the implementation of specific control and mitigation measures, liaison with external bodies and the public, and measures for clean-up and restoration.
- Training and drills are important to ensure everyone is familiar with the emergency plan and it is implemented promptly and effectively.

Recognising Risk Situations and Action to Take

It is important that, as part of the training of employees, they become familiar with what is an emergency and what action to take. Not all spillages will be emergencies, although they will all likely need to be dealt with and cleaned up:

- A small spill of oil in a bunded area is unlikely to be classed as an emergency, but a spill of oil in an uncontained area with the potential to pollute a river needs to be treated as an emergency and dealt with immediately.
- A spill of diesel in an area that can be closed off to traffic need not necessarily be an emergency, but if it were petrol instead of diesel, on a hot day with possible ignition sources nearby, the situation would be very different.

4.2 Emergency Preparedness and Response

> **TOPIC FOCUS**
>
> **COMAH**
>
> **Directive 12/18/EU** on the control of major accident hazards involving dangerous substances is generally referred to as the **COMAH Directive**. It is covered in the UK by the **Control of Major Accident Hazards Regulations 2015**.
>
> The **COMAH Directive** was introduced in the wake of the Seveso disaster. This emergency occurred in 1976 at a chemical plant located close to the Italian town of the same name. A process failure at the plant led to the release into the atmosphere of quantities of a highly toxic dioxin compound. Several hundred people in the surrounding area suffered skin lesions and over 80,000 farm animals had to be slaughtered to prevent dioxin entering the food chain.
>
> The aim of the Directive is to reduce the risk of similar accidents occurring at installations that handle significant quantities of hazardous substances:
>
> - The **COMAH Directive and Regulations** define upper and lower threshold quantities for a specified range of hazardous chemicals.
> - Any site that stores any of the specified chemicals in quantities greater than either of the threshold quantities, must register with the Competent Authority (the Health and Safety Executive (HSE) and Environment Agency in England; the HSE and Natural Resources Wales in Wales; the HSE and Scottish Environment Protection Agency in Scotland).
> - If only the lower threshold quantity is exceeded, the site is designated as a 'lower-tier' site; any site which exceeds the higher threshold becomes an 'upper-tier' site.
>
> Designated sites must take a range of actions to prevent accidents and limit their consequences, such as:
>
> - Provide basic site details to the Competent Authority.
> - Prepare a major accident prevention policy.
>
> Upper-tier sites must also:
>
> - Produce a regular site-safety report.
> - Prepare and test an on-site (internal) emergency plan.
> - Supply information to the local authority to enable an off-site (external) emergency plan to be prepared.
> - Provide information to local residents about the substances held and emergency arrangements.

Any unplanned fire should be treated as an emergency, as there is great potential for fire to spread quickly, causing great damage and cutting off potential evacuation routes. However, if the fire is within the waste of a landfill site, it is unlikely the fire service will be called, unless there is a risk of the fire spreading beyond the waste area to buildings. This is because landfill site staff will be trained to deal with this situation, and will have equipment that can reach the point of the fire, whereas it is unlikely that the fire service will be able to get their equipment across the rough ground of most landfill sites.

Ensuring employees respond at the correct level of urgency will help to make sure that incidents are dealt with effectively, and also prevent the development of a culture where any minor spill is treated as a major emergency. If this happens, the danger is that when a major emergency does occur, people feel it is just another case of overreacting, possibly leading to an inappropriate response.

Many environmental impacts are associated with emergency situations. A number of severe examples were given in Element 1, such as the release of toxic gas into the atmosphere at Bhopal in India in 1984. Incidents like this are very rare, however, and hopefully will never be experienced by most organisations. But smaller-scale incidents, such as spills of oil and minor fires, are much more common and so all organisations have a responsibility to consider emergency events in their environmental management planning to prevent or mitigate environmental damage.

Integration of Environmental Risk in Emergency Plans and Procedures

Emergency plans will cover many issues such as protection of workers, local community, business continuity and environmental receptors. An organisation will not have separate plans covering each of these issues - there is one overall emergency plan. It is important that such a plan considers the implications of environmental damage caused by an emergency.

Environmental Hazards Associated with Fire

The environmental hazards associated with fire will, to a large extent, depend on what is actually on fire. However, there are some hazards common to almost all fires:

- **Air Pollution**

    From any accidental fire there will always be smoke; the size of the fire and the properties of the burning material will determine how serious any air pollution is and how far that pollution will spread. Air pollution will also be impacted by weather conditions that may reduce or increase the severity of the pollution. Care should be taken during the preparation of the emergency plan to identify any potential receptors that may be particularly susceptible to the type of air pollution that would be caused by an accidental fire from your organisation. Potential receptors may be:

    - Natural environments easily damaged by poor air quality.
    - Human environments, such as care homes or hospitals.

    Suitable advice for these receptors can be prepared in advance, and consideration given to including the management of these facilities in some of the training and practice drills undertaken.

- **Water Pollution**

    Surface water is easily polluted by both firewater run-off, and spills of oil or chemicals that occur as a result of the fire. It is essential that:

    - Sensitive water receptors are identified and noted in the emergency plan.
    - Procedures are in place to ensure potentially damaging substances are prevented from reaching these receptors. In some cases, this may require holding tanks or other specific areas to store the run-off water so that it can be treated before being discharged to the watercourse.

- **Land Pollution**

    Land may become contaminated in exactly the same way as water does (see above). If land is used for a sensitive use such as agriculture or, for example, it is a Site of Special Scientific Interest (SSSI) then it is important that it is protected. It is also imperative to understand what lies underneath the land as this may provide a pathway for future pollution of groundwater.

Emergency Response Plans

Organisations of all types and sizes need to have an emergency plan in place. The key point is that the emergency plan should be appropriate to the scale and nature of the organisation's activities. A major facility, such as an oil refinery, will clearly have a much more extensive and detailed emergency plan than an office-based service organisation. In fact, larger industrial organisations that store chemicals above certain threshold limits (e.g. greater than 50,000 litres of fuel) are required, under **EU Directive 12/18/EU** on the control of major accident hazards involving dangerous substances (known as the **COMAH Directive**) (covered in the UK by the **Control of Major Accident Hazards Regulations 2015**); (covered in the UK by the Control of Major Accident Hazards Regulations 2015), to have an emergency plan in place that meets specified criteria.

4.2 Emergency Preparedness and Response

Environmental Content of Emergency Plans

Emergency plans should include foreseeable internal and external causes such as the following.

Internal

- **Fire**

    Fires may result in hazardous materials being released into the atmosphere, or washed into watercourses by fire hoses being used to extinguish the fire.

- **Spillages**

    Many organisations store fuel and other hazardous liquids, such as chemicals. Losses of hazardous liquids can occur in circumstances such as:

    - When a tanker is delivering fuel and the delivery hose leaks, or the receiving tank overflows.
    - Decanting a liquid from one container to another by hand.
    - If a storage tank or pipeline becomes corroded and leaks.
    - A vehicle, such as a forklift truck, collides with tanks or drums.

- **Plant Failure**

    If pollution abatement equipment breaks down or malfunctions, emissions and discharges may fail to meet regulatory limits.

- **Loss of Containment**

A sudden loss of containment from pipework, pumps and storage or process vessels or other forms of containment is a major cause of accidents.

External

Extreme weather events - for example flooding may cause flood waters flowing across a site to wash hazardous materials onto adjacent land or into local watercourses.

Materials to Deal with Spills

It is important that the correct equipment is available to deal with the spillages that are likely to take place. Suitable spill kits need to be located at key locations around the site. This will include locations near to:

- Where chemicals are stored.
- Any drainage points that may act as pathways for the pollutant.

It is essential that the spill kits contain the correct materials for the possible pollutants and that people are trained in their use. Colour-coding is used to identify the type of pollutant that the material can deal with. For example, some pads soak up oil but leave water behind, so it is important that the correct pads are used.

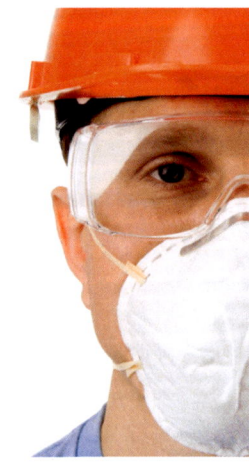

> **TOPIC FOCUS**
>
> **Common equipment contained in a spill kit might include:**
>
> - Mats for soaking up spills.
> - Socks or booms for containment.
> - Drain covers to seal drains and prevent pollution entering the drainage system.
> - Granules for soaking up oils.
> - Gloves and other relevant PPE.
> - Wipes.
> - Bags to put contaminated materials in.

It is important that spill kits are stored in suitable containers, especially if the kits are to be stored outside.

Emergency Control Centre (ECC)

Larger sites usually have a separate, dedicated on-site building called the Emergency Control Centre (ECC). It should be located sufficiently far away from foreseeable sources of major incidents so that it remains operational throughout.

The ECC acts as the nerve centre for operational management by the main controller during an on-site emergency. It is a prime consideration in the emergency plan. The ECC would typically be stocked with:

- On-site and off-site communication equipment (telephones, two-way radio).
- Site plans/maps (marked with locations of site hazards, safety equipment, drainage routes, firewater supplies, fire-fighting equipment, local features, etc.).
- Contact details (including those for emergency response personnel, head office, hospitals, regulators, emergency services, media - though dealing with the media may be delegated to off-site personnel).
- Facilities to record the development of the emergency (such as whiteboards, plans and maps).
- Facilities to access data on those present at the facility at the time of the incident (i.e. roll-call).

It will probably not be feasible, or appropriate, for smaller organisations to have a dedicated ECC. Nevertheless, every organisation should have a central point for holding key information about the emergency plan and from where incident responses can be co-ordinated. This might be the gatehouse, or a senior manager's office. The location should be protected, as far as possible, from the effects of any likely accident, e.g. by being close to the site perimeter.

Information and Training for Internal and External Interested Parties

Training and Practices

Emergencies happen when we least expect them and, during an emergency, situations can change very quickly. For these reasons:

- It is essential that any employees or contractors who may be involved in dealing with an emergency situation are fully trained in their roles and responsibilities.
- Training may involve local emergency services, such as the fire service, ambulance or police. This will help to ensure that, should they be required during an emergency, they will have a good idea of the layout of any buildings as well as any potential hazards on site.

4.2 Emergency Preparedness and Response

Testing should be based on an accident scenario identified as being reasonably foreseeable. Tests should address the response during the initial emergency phase, which is usually the first few hours after the accident occurs. This is the phase of an accident response when key decisions, which will greatly affect the success of any mitigation measures, must be made under considerable pressure and within a short period of time. This is, therefore, where a detailed understanding of the likely sequence of events and appropriate countermeasures is of great benefit.

The objective of testing the emergency plan should be to give confidence in the following constituents of the plan:

- Completeness, consistency and accuracy of the emergency plan and other documentation used by organisations responding to an emergency.
- Adequacy of the equipment and facilities, and their operability, especially under emergency conditions.
- Competence of staff to carry out the duties identified for them in the plan, and their use of the equipment and facilities.

The training does not always need to take the form of an exercise where potential accident situations are simulated, although carrying out these simulation exercises on a regular basis can:

- Be a great help in identifying problems and solutions.
- Give the participants an indication of how they will behave during an incident.

Other training methods can be used, such as tabletop exercises. These allow information exchange and dissemination between the organisations that may be involved in an incident, as well as decision-making, to be tested. They can be carried out in relation to a model, plans or photographs to depict the establishment.

Access to Site Plans

In order for the emergency plan to be effective, information about the site needs to be compiled and shown on a drawing easily available in the event of an emergency. Information should include:

- The layout of the site drainage system showing clearly the difference between drains going to foul sewer and those going to surface water.
- Assembly points for staff and visitors.
- Access routes and assembly points for emergency services.
- The location of any fire hydrants or high-pressure water points.
- The location of any flammable or explosive chemicals stored on site and any other locations that may prove dangerous to emergency service personnel.

It may be necessary to keep a number of copies of all these plans to ensure that at least one will always be available. It may also be advisable to locate a copy of the drawing in a cabinet at the main entrance to the site and/or the Emergency Control Centre; this will help to ensure that, should an incident occur when the site is unoccupied, emergency services still have access to the information before members of staff that may be on call are able to arrive at the site.

Inventory of Materials

A key piece of information required in planning for any major emergency is the quantities of dangerous substances held (or likely to be held) on site at the time of the incident. In combination with knowledge of the nature of the material, this enables an estimate of the likely:

- scale,
- extent; and
- severity

of any incident. This information is needed for the:

- Development of on-site emergency plans.
- Emergency services so that they can plan for and mount an effective off-site response.

Quantities of dangerous substances on site will need to be known

Liaison with Regulatory Bodies and Emergency Services

It is good practice, even for smaller organisations, to discuss potential emergency situations with the emergency services and the environmental regulator, to identify the best response approaches. Should an incident occur, the emergency services will also then have some prior knowledge of the layout of the site and the conditions they are likely to encounter.

The **COMAH Directive/Regulations** set out requirements for communicating and liaising with regulating bodies and emergency services. They also require competent authorities to develop their own plans for dealing with a major incident. For those organisations not covered by national laws made to comply with the **COMAH Directive/Regulations**, it is still good practice to discuss potential emergency situations with the relevant enforcement authorities and emergency services to establish the best ways to resolve them.

If an emergency has occurred where there is any potential for a pollution incident to occur, e.g. chemicals have been spilt that might reach a local watercourse, there is a responsibility on the organisation to inform the regulator as soon as possible.

In some parts of the world, fire and rescue services have the ability to store and access plans and specific information while on route to an incident. Regular communication with the local fire and rescue service will ensure they have accurate, up-to-date information that they can rely on upon arrival at the scene. This helps to ensure that they can react quickly and safely to deal with the incident and that any specialist equipment required can be called out as early as possible.

Protecting and Liaising with Local Residents

Local residents (including any indigenous peoples) need to be kept informed about any incidents that might affect them. The emergency services may play an important role in this process, especially if any evacuation is required. The management of these communications should be included in the emergency plan. If people are not kept properly informed, then an accident could severely affect their health and safety and spread panic. Important methods of communication jammed by excessive use from neighbours and members of the public seeking information about the incident, hinder the effectiveness of the emergency response.

Handling the Press and Other Media

Dealing with the press and other media is often an issue that is not given full consideration until an emergency incident takes place and it then becomes clear that there is significant interest from local and possibly national media organisations. The reputation of an organisation is arguably one of its most important assets and one that is easily lost or damaged through poor communication to the public via the media.

In the absence of any accurate, open and honest information coming from the organisation, it is potentially inaccurate information that is publicised. Once that information is in the public domain, the organisation is in the position of having to deny or correct the information.

During the planning process for emergencies, it is important that:

- Assigned spokespersons are identified as being those who will communicate with the media during an emergency.
- All other employees are given clear instruction not to talk to reporters and to direct any enquiries to those members of staff who have been allocated this role.
- Nominated staff members are competent in how best to communicate with the media.
- When an emergency occurs, there are systems and procedures in place to ensure that nominated staff are kept fully informed of the situation regarding the emergency.

Building good relationships with local press, radio and television reporters can be very beneficial, should an emergency incident occur, as they will automatically approach the member of staff they normally deal with, who is likely to be someone trained and experienced in dealing with the media. Local media can also be supportive in getting important mitigation information to the local population, such as the need to stay indoors and close all windows. This information will be more effectively distributed if the different parties involved are used to dealing with each other.

4.2 Emergency Preparedness and Response

Review and Continual Improvement of Emergency Response Plans

It is important to review the lessons learned from testing, to determine whether modifications are required to the emergency plan. With the different organisations involved in emergency plan tests, there will be more than one method for evaluating the effectiveness of the emergency plan, and each organisation may want to establish its own self-evaluation criteria relevant to its own response. For example, organisations may want to set quantitative measures like timeliness of response, or subjective measures for quality of performance.

This evaluation process needs to include the dissemination of information and the lessons learned, as appropriate, to the relevant response organisations, who need to be kept informed of progress on any actions to amend emergency plan responses. This will also cover any recommendations arising from the testing and the progress of actions to maintain an effective plan.

> **STUDY QUESTIONS**
>
> 2. What is the main objective of testing an emergency plan?
> 3. List three pieces of information about the site that should be included in an emergency plan.
> 4. List the three main environmental hazards associated with fires.
>
> (Suggested Answers are at the end.)

Summary

This element has dealt with environmental emergencies and how to plan for and deal with them.

In particular, this element has:

- Demonstrated that, as part of their general duty or responsibility not to pollute, all organisations should have emergency plans in place, with the size and complexity of the plan depending on the nature of the organisation.
- Shown that emergency plans are necessary:
    - As part of the EMS - potential impacts under emergency conditions.
    - To ensure prompt action to protect people and mitigate potential consequences to the environment.
    - To pre-empt the possibility of prosecution and costs relating to fines, clean-up work and compensation.
    - To avoid adverse publicity in the event of an incident occurring.
- Outlined how the **COMAH Directive/Regulations** require the preparation of emergency plans.
- Explained that on large sites the Emergency Control Centre acts as the nerve centre for operational management during an on-site emergency and should contain communication equipment, site plans, contact details, information recording facilities, etc.
- Shown that training in emergency procedures, including the use of simulation exercises, should be given to all employees/contractors. Training in the recognition of significant/less significant incidents should be included.
- Explained that there should be appropriate spill kits, stored in suitable containers, and that full site plans should be available as well as an accurate inventory of any dangerous substances on site.
- Outlined the environmental hazards associated with fire, including:
    - Air pollution: from smoke.
    - Water pollution: from firewater run-off; spills of oil or chemicals.
    - Land pollution: as per water pollution.
- Explained that regular communication with regulatory bodies and the emergency services will help to ensure efficient and speedy action when required by the occurrence of an emergency incident. The establishment of good relationships with the local media may prove to be beneficial to the public image of the organisation in the event of an incident occurring.

Exam Skills

Question

Scenario

Following the results of an audit from an external consultant, you find the emergency plan for your site is poor. You notice that although there are some effective procedures for dealing with an emergency, they are not comprehensive and have not been fully implemented.

Task

What are the actions that are required to ensure that the emergency plan will work effectively when needed?

(8 marks)

Approaching the Question

Think about the steps you would take to answer the question:

- Read the scenario carefully. With this question you need to develop actions that will mean that the emergency plan will work properly during an emergency situation. Note: the question does not cover the development of an emergency plan or its content.
- Now look at the task – prepare notes on:
  - Measures that could be taken to ensure that the emergency plan is put into practice in an effective manner.
- Consider the marks available. In this case, there are 8 marks available so you should provide at least 8 measures that could be implemented.
- Read the scenario and task again to make sure you understand them and have a clear understanding of emergency plans and their implementation.
- Jot down an outline plan - this might include:
  - Training, practices/drills, past incident analysis, review of plan, maintenance of systems/alarms, copies held at relevant locations, Emergency Control Centre equipment, audit of plan.

Now have a go at the question yourself.

Exam Skills

Example of How the Question Could be Answered

A key element for the effective operation of an emergency plan is that key staff are trained as to its contents. This should include both training for new staff and refresher training for existing staff.

The organisation should also undertake practices and drills - this may involve neighbours and external responders to the incident, such as the fire authority if appropriate.

If incidents do occur, they need to be closely analysed. If weaknesses in the plan are identified then the plan may need to be altered to ensure such shortcomings do not recur.

The plan should also be reviewed on a frequent basis and as the situation changes. The plan should reflect the actual emergency incidents which are likely to occur at the site.

Any systems and alarms should be regularly maintained to ensure that they are working correctly. Equipment that should be subject to regular maintenance includes emergency signage, visitor instructions and personal protective equipment.

Copies of the plan should be available at relevant locations. For example, a copy should be held in the Emergency Control Centre (ECC) which should be clearly identified as such. A copy of the emergency plan should also be held in an off-site location.

The ECC should also contain relevant equipment, such as communication devices, and contact details for relevant internal and external persons with regard to the plan.

The plan should also be audited on a regular basis (it may form part of a formal EMS such as those designed to ISO 14001). External audits by an environmental specialist will give a high level of confidence in the plan.

Element 5

Control of Emissions to Air

Learning Outcomes

- Understand the importance of reducing environmental harm; identify sources of air pollution; and suggest suitable control measures.

Learning Objectives

Once you've read this element, you'll be able to:

1. Demonstrate awareness of the environmental impacts of air pollution.

2. Identify sources of environmental harm and suggest suitable control measures for emissions.

Contents

Air Quality Standards — 5-3

Why are there Air Quality Standards? — 5-3
Meaning and Uses of ppm and mg/m^3 — 5-3
The Potential Effects of Poor Air Quality — 5-3
The Role of Air Quality Standards — 5-4

Main Types of Emissions to Atmosphere — 5-5

Types of Emission and their Hazards — 5-5
Sources of Air Pollution — 5-6
Common Pollutants — 5-8

Control Measures to Reduce Emissions — 5-12

Controlling Air Pollution — 5-12
Control Hierarchy — 5-12
Examples of Technology — 5-13
Unit EMC2: Environmental Practical Application — 5-25

Summary — 5-26

Exam Skills — 5-27

Air Quality Standards

IN THIS SECTION...

- Air pollutants are measured in parts per million (ppm) or milligrams per cubic metre (mg/m³).
- There are a number of health effects associated with poor air quality. These include short-term irritation and inflammation, long-term respiratory problems and, in some cases, an increased incidence of cancers.
- Standards of air quality are set under the European Union **Directive 2008/50/EC** on standards of air quality for certain pollutants. The aim is to protect the environment and human health.

Why are there Air Quality Standards?

The quality of the air that surrounds us is critical to health and the enjoyment of life. Poor air quality may also have detrimental impacts on wildlife and agricultural systems. But pollution of the air we breathe has been a problem for centuries, as cities have grown larger and human society has become more industrialised. Although progress has been made in controlling some sources of air pollution, air quality is still declining in many areas of the world, especially in large cities with high densities of road vehicles. Air quality standards are an important tool for regulators in controlling levels of the atmospheric pollutants to which we are potentially exposed.

Meaning and Uses of ppm and mg/m³

Levels of contaminants in the atmosphere are typically expressed in units of either 'ppm', or 'mg/m³'; it is important that you clearly understand the differences between these measures.

The unit 'ppm' is shorthand for parts-per-million. This is a way of denoting the relative proportion of a contaminant in a sample of air, usually on a volume basis. For example, if a sample of air of one litre in volume contains one-millionth of a litre of a pollutant, then the level of the pollutant may be expressed as 1ppm. The unit ppm is typically (although not always) used for pollutants that exist in the atmosphere as gases or vapours.

The unit mg/m³ expresses the concentration of a pollutant in terms of the mass (milligrams) of the substance present in a given volume (one cubic metre) of air. For example, if a sample of 1m³ of air is found to contain 10mg of dust particles, then the concentration of dust may be expressed as 10mg/m³. The unit mg/m³ is typically used for contaminants that exist in particulate form.

The Potential Effects of Poor Air Quality

We have seen (Element 1) that pollution may have local, regional and global impacts. This is especially relevant when considering the potential impacts of air pollution because of the ability of some contaminants to be transported significant distances through the atmosphere.

Local air pollution, especially in towns and cities, is a continuing focus for regulators because of the health impacts of ground-level pollution, such as city smog. The World Health Organization has estimated that there are about 3 million deaths annually worldwide associated with poor local air quality. In many developed countries local air quality has improved markedly with the introduction of stringent air pollution legislation. Nevertheless, problems remain. For example, the Royal College of Physicians and the Royal College of Paediatrics and Child Health have reported that air pollution contributes to 40,000 deaths a year in the UK.

Poor air quality

Short-term human health effects from air pollution can include irritation and inflammation of the airways, eyes and mouth, and triggering of asthma attacks in susceptible individuals. Longer-term problems are often associated with cardio-pulmonary (heart-lung) performance, but some pollutants are also associated with increased incidence of cancers.

5.1 Air Quality Standards

> **MORE...**
>
> For more information about the effects of poor air quality on human health, see the Healthy Air Campaign at:
>
> http://healthyair.org.uk

The long-range transport of air pollution from cities and industrial centres can also have regional impacts. Oxides of nitrogen (NO_x) and sulphur (SO_x) can travel hundreds of kilometres from their emission source and react with moisture in the atmosphere to create 'acid rain'. This affects trees by damaging their leaves and bark, making them more vulnerable to disease, weather and insects. Toxic amounts of aluminium and iron may also be released from soils, further damaging trees and other plants. Lake ecosystems that are exposed to acid rain may become acidic and this can kill fish eggs. At higher levels of acidity, aluminium may build up in the water and kill adult fish, which are important food sources for other animals such as birds.

We have also seen (Element 1) how global effects of air pollution - climate change and ozone depletion - have now moved to the top of the international agenda.

The Role of Air Quality Standards

Objectives and standards for the quality of the air that surrounds us form an important element of both international and national environmental policy. Objectives and standards are usually set in the form of maximum concentrations of specified pollutants that should not be exceeded in the local atmosphere. In the EU, **Directive 2008/50/EC** sets target limits for the maximum levels of a range of pollutants: fine particles, sulphur dioxide, nitrogen dioxide, lead, carbon monoxide, benzene, ozone, arsenic, cadmium, nickel and Polycyclic Aromatic Hydrocarbons (PAHs). The Directive as it applies to England has been covered by the **Air Quality Standards Regulations 2010 (as amended)** with similar legislation developed in other parts of the UK. The Directive as it applies to England has been covered by the **Air Quality Standards Regulations 2010 (as amended)** with similar legislation developed in other parts of the UK.

The emission limits that are imposed via permits on industrial installations (for example, under **Directive 2008/1/EC** on integrated pollution prevention and control), as well as other controls - for example, on vehicle engine emissions - are intended to ensure that these ambient air-quality standards are not breached.

The long-range movement of air pollutants and regional impacts are the subject of other international agreements, most notably the Geneva Convention on Long-Range Transboundary Air Pollution (and eight associated Protocols). This aims to limit and gradually reduce pollutants that can cross national boundaries. The Convention applies to the countries within the United Nations Economic Commission for Europe (UNECE) - this includes 56 member states across Europe and also includes Canada, the USA, Central Asian republics and Israel.

> **MORE...**
>
> The World Health Organization (WHO) publishes a set of air-quality guidelines for various pollutants. These guidelines are available at:
>
> www.who.int/airpollution/guidelines/en
>
> The *Air pollutant emission inventory guidebook* provides guidance on estimating natural and anthropogenic emissions to help in reporting emissions as required by the Geneva Convention on Long-Range Transboundary Air Pollution. It is available at:
>
> www.eea.europa.eu/themes/air/emep-eea-air-pollutant-emission-inventory-guidebook/emep

STUDY QUESTION

1. What can be the short-term effects of human exposure to air pollution?

(Suggested Answer is at the end.)

Main Types of Emissions to Atmosphere

IN THIS SECTION...

- There are various forms of air pollutants, including gases, vapours, mists, particles, smoke, dust, grit, fibres, odours and fugitive emissions.
- Fossil fuels, industrial processes and transport are all sources of air pollution.
- Common air pollutants include sulphur compounds, nitrogen compounds, halogens and their compounds, metals and Volatile Organic Compounds (VOCs).

Types of Emission and their Hazards

The contaminants that create air pollution may exist in a variety of forms: gases, vapours, mists and particles.

Gaseous

Substances that remain in the gaseous phase at normal temperatures and pressures, such as carbon dioxide, nitrogen and ozone.

Liquids Suspended in Air

- **Vapour**

  Gaseous state of materials that are liquid at normal temperature and pressure (e.g. steam).

- **Mist**

  Mists are fine liquid droplets suspended in the air, usually nucleated by a particle.

The health implications of liquids suspended in the air are largely dependent on the properties of the substance involved; oil or acid mists, for example, may cause irritation of the respiratory system.

There are health implications with emission to atmosphere

Particles

Particles consist of solid matter of all sizes, from less than 0.001 microns to greater than 100 microns (1 micron is 1 millionth of a metre).

- **Smoke**

  Particles in the range 0.1 microns to 10 microns are seen as smoke. There are no clearly established size definitions for these particulates and different publications suggest other overlapping size bands.

- **Dust**

  A dust may consist of any size or shape of particle, crystalline or amorphous.

- **Grit**

  This is a general term for coarse, solid particulate matter and is defined as solid particles greater than 75 microns.

- **Fume**

  This refers to very small particles, of less than 1 micron, that are suspended in flue gases and air.

Particle sizes capable of being inhaled are up to 10 microns; particle sizes of less than 7 microns are capable of penetrating lung tissue.

Such particles clearly represent at least an inhalation hazard for humans and animals and, depending on their chemical nature, may lead to acute effects (such as irritation of eyes, nose, throat) or longer-term ill health (such as asthma). Particulates, together with odours, can also represent a significant nuisance, interfering with people's enjoyment of their surroundings. We looked at some other general issues associated with airborne particulates in Element 1, when we considered the size of the environmental problem.

Fibre

Fibres are particles given off from materials of a fibrous nature, such as wood and asbestos. Fibres will usually have an elongated shape and so cannot be compared to other particles in terms of size, as they may be very narrow but long particles.

The effects of inhalation of fibres depend largely on the type of fibre. For example, inhaling asbestos or silica can cause inflammation and scarring in the lungs - this is commonly known as asbestosis or silicosis, depending on the substance involved. Asbestos exposure may also cause changes to the lining of the chest cavity (pleura) and increase the risk of lung cancer and mesothelioma (cancer of the pleura).

Odours

We referred to the issue of unpleasant odours in Element 1, where we noted that these can cause a nuisance to local residents.

Fugitive Emissions

These are emissions that should not be there, i.e. they have escaped from a process in a manner that is unplanned, such as leaks from pipework or contained conveyor systems. They are often difficult to quantify owing to their dispersed nature.

Sources of Air Pollution

> **DEFINITION**
>
> **FOSSIL FUEL**
>
> A natural fuel, such as coal or gas, formed from the remains of living organisms that died millions of years ago and became fossilised beneath the Earth's surface.

Fossil Fuels

As our population grows, so does our consumption of goods and services. Our energy consumption also increases to meet the growing demand. Currently, most of the energy we use comes from burning fossil fuels, so an increase in energy consumption tends to result in increased air pollution.

The most common fossil fuels are:

- Coal (including anthracite and lignite).
- Coke.
- Petrol/gasoline.
- Diesel/DERV.
- Fuel oil/heating oil/paraffin.
- Natural gas (methane).
- Liquefied Natural Gas (LNG).
- Liquefied Petroleum Gas (LPG), propane or butane, or a mixture.

Different types of fuel have different sources - coal comes from the remains of trees and ferns, whereas natural gas and crude oil tend to come from dead marine plants. As the vegetation died it sank; the pressure above it - from layers of sediment - caused it to gradually transform; and the carbon from the plants was compressed into fossil fuel.

Main Types of Emissions to Atmosphere | 5.2

All fossil fuels have a very high carbon content. This carbon reacts with oxygen in the atmosphere when they are burnt. This reaction releases a large amount of carbon dioxide into the atmosphere - which makes a significant contribution to climate change. It also releases a large amount of energy in the form of heat and light.

In addition to carbon dioxide, many other contaminants can also be released into the environment - depending on the chemical make-up of the fossil fuel and any additives or impurities which may be present (e.g. lead in petrol). The most common contaminants released when fossil fuels are burned include:

Contaminant	Description
Carbon dioxide	One of the main greenhouse gases; contributes to climate change.
Sulphur oxides (SO_x)	Contributes to the formation of acid rain.
Nitrogen oxides (NO_x)	Contributes to the formation of acid rain and city smog.
Particulates	Small particles of soot and ash; may cause respiratory problems.

In addition to these common contaminants, small quantities of highly toxic substances can also be released. These can include radioactive substances and compounds of mercury and lead.

Industrial Processes

Industrial activities almost all use fossil fuels as a power source and, as a result, contribute to air pollution.

In addition, some industrial processes release additional harmful pollutants into the atmosphere. The main industrial sources of air pollutants are:

Industry	Air Pollutants
Oil and gas refining	Particulates, sulphur oxides (SO_x), nitrogen oxides (NO_x), Volatile Organic Compounds (solvent fumes), benzene, hydrogen sulphide, hydrogen fluoride.
Cement manufacturing	Particulates, carbon dioxide, sulphur oxides (SO_x), nitrogen oxides (NO_x), hydrogen chloride, benzene, toluene, xylene.
Metal smelting and refining	Particulates, carbon dioxide, sulphur oxides (SO_x), nitrogen oxides (NO_x), hydrocarbons, acid gases.
Pulp and paper manufacturing	Particulates, Volatile Organic Compounds (solvent fumes), chloroform, formaldehyde, ammonia.
Chemical production	Volatile Organic Compounds (solvent fumes), acid gases.
Waste incineration	Particulates, carbon dioxide, sulphur oxides (SO_x), nitrogen oxides (NO_x), hydrogen chloride, hydrogen fluoride.

Transport

Motorised road transport has grown almost exponentially over the last 50 years. Around half of the world's oil production is now consumed by cars, trucks and buses. There has been a similar expansion of air travel, which also depends entirely on fossil fuel.

As a result, fossil-fuel combustion for transportation is now a major source of air pollution, especially poor air quality in city centres. It has been estimated that in industrialised countries the transport sector accounts for 70-90% of all carbon monoxide emissions, 60% of particulate emissions and 40-70% of nitrogen oxide (NO_x) emissions. Until quite recently, the compound tetraethyl lead was widely used as an additive in petrol and this was the source of high levels of lead in the atmosphere and urban soils which has been linked with neurotoxic effects, especially in children.

Common Pollutants

Sulphur Compounds

> **TOPIC FOCUS**
>
> **Sulphur Compounds**
>
> Man-made sources:
>
> - Combustion of fossil fuels, e.g. coal and oil. Combustion may be in industrial processes, domestic fires, boilers and vehicles.
>
> Effects:
>
> - Acid rain.
> - Irritant to the eyes, nose and lungs. Irritation to the lungs can cause respiratory problems.

Compounds of sulphur that form atmospheric pollutants include:

- Sulphur dioxide (SO_2).
- Sulphur trioxide (SO_3).
- Sulphuric acid.
- Sulphate salts.
- Hydrogen sulphide.

Much of the atmospheric sulphur comes from natural sources such as volcanic activity, decaying organic matter, phytoplankton (vegetable plankton) and sulphur springs. These natural sources may account for up to half the total volume of current atmospheric sulphur. Man-made sources will account for the remaining atmospheric sulphur.

Nitrogen Compounds

> **TOPIC FOCUS**
>
> **Nitrogen Compounds**
>
> Man-made sources:
>
> - Combustion of fossil fuels, e.g. coal and oil. Combustion may be in industrial processes, domestic fires, boilers and vehicles.
>
> Effects:
>
> - Acid rain.
> - Depletion of the ozone layer.
> - Generation of photochemical smogs.

The main compounds of nitrogen in the atmosphere are:

- Nitric oxide (NO).
- Nitrogen dioxide (NO_2).

Nitrogen is a commonly occurring natural element and indeed makes up almost 80% of the atmosphere. The main man-made sources of oxides of nitrogen are similar to those for sulphur, being mainly combustion processes. Little nitric oxide (NO) is produced at room temperature, but with increasing temperature more NO is produced.

While NO has no known adverse health effects, it is slowly converted into NO_2 in the atmosphere, and NO_2 is known to be toxic. NO_2 forms secondary pollutants, such as tropospheric ozone (O_3), which is a strong oxidising agent and highly reactive.

Halogens and their Compounds

Halogens are a group of elements such as fluorine, chlorine, bromine and iodine. Fluorine and chlorine are gases in their natural state, while bromine is a liquid and iodine a solid. They are highly reactive elements and can therefore be harmful, or lethal to biological organisms in sufficient quantities. Halogens occur in numerous compounds.

> **TOPIC FOCUS**
>
> **Halogens and their Compounds**
>
> - Man-made sources:
> - Disinfectants in drinking/swimming water (chlorine, bromine).
> - Toothpaste additive (fluorine).
> - Treatment of materials against fire damage, as they are fire-retardant (e.g. halons - bromine).
> - Chlorofluorocarbons (CFCs) (see Element 1).
> - Pesticides, herbicides and fungicides (e.g. DDT - chlorine).
> - Dyes, soaps and medicines (iodine).
>
> Effects:
>
> - Attack inert materials such as glass (particularly fluorine).
> - Burns the skin and a respiratory irritant.
> - Fatal in high doses (particularly chlorine).
> - Ozone depletion.

5.2 Main Types of Emissions to Atmosphere

Metals

The main metal pollutant in the air is from lead that was added to petrol to improve engine performance. When leaded petrol is burnt, lead is emitted from the exhaust, and concern has focused on health effects. Lead emissions from road vehicles are falling as a result of the use of unleaded petrol.

> **TOPIC FOCUS**
>
> **Lead**
>
> Man-made sources:
>
> - Leaded fuel (we now use unleaded).
> - Lead-based paints.
> - Lead water pipes.
> - Metal industry.
>
> Effects:
>
> - High levels - nausea, vomiting, convulsions, death.
> - Low levels over long periods:
>   - Irritability, sleeplessness, fatigue.
>   - Constipation, headache, loss of appetite.
>   - Anaemia.
>   - Damage to the brain and nervous system.
>   - Kidney damage.

Volatile Organic Compounds (VOCs)

Volatile Organic Compounds (VOCs) are gases emitted from certain solids and liquids. They easily vaporise at room temperature. There are many different VOCs; examples include benzene, toluene and Polycyclic Aromatic Hydrocarbons (PAHs).

> **TOPIC FOCUS**
>
> **Volatile Organic Compounds (VOCs)**
>
> Man-made sources:
>
> - Exhaust fumes.
> - Cigarette smoke.
> - Synthetic materials.
> - Household chemicals.
> - Paints.
> - Glues and adhesives.
>
> Effects:
>
> - Involved in formation of ground-level ozone and photochemical smog.
> - Eye, nose and throat irritation.
> - Dizziness and memory impairment.
> - Some VOCs cause cancer.

Greenhouse Gases

The key greenhouse gas is carbon dioxide. Carbon dioxide is naturally emitted from sources such as volcanic activity but is also emitted in massive quantities from human activities such as the burning of fossil fuels.

Some of the key greenhouse gases and their sources are identified in the table below:

Name	Main Source
Carbon dioxide (CO_2)	Combustion of fossil fuels, deforestation
Methane (CH_4)	Management of wastes, agricultural activities (e.g. rice production)
Nitrous oxide (N_2O)	Agricultural soils, combustion of biomass
Hydrofluorocarbons (HFCs)	Refrigeration
Perfluorocarbons (PFCs)	Industrial activities, e.g. electronics and aluminium production
Sulphur hexafluoride (SF_6)	Industrial activities, e.g. semiconductor manufacture

Global Warming Potentials (GWPs) are a way of measuring the strength of various greenhouse gases. The GWPs provided in the table below are considered over three time horizons (20 years, 100 years and 500 years). Some greenhouse gases are very gradually removed from the atmosphere and it can take much longer than 20 years. The GWP of carbon dioxide is 1 and all greenhouse gases are compared to this level.

Greenhouse Gas	Global Warming Potential (GWP)		
	Time Horizon		
	20 years	100 years	500 years
Carbon dioxide	1	1	1
Methane	62	23	7
Nitrous oxide	275	296	156
CFC-12	7900	8500	4200
HCFC-22	4300	1700	520

As we can see, although CO_2 in its own right has a low GWP in comparison to other gases, it is the quantity that is emitted that leads to it being the most important greenhouse gas.

DEFINITION

GLOBAL WARMING POTENTIAL

A measure of how much heat a greenhouse gas traps into the atmosphere relative to the heat trapped by a unit of CO_2.

STUDY QUESTION

2. Define the terms 'vapour', 'mist' and 'fume'.

(Suggested Answer is at the end.)

5.3 Control Measures to Reduce Emissions

Control Measures to Reduce Emissions

IN THIS SECTION...
- There is a control hierarchy to reduce emissions, with elimination being the preferred option, followed by minimisation and then finally rendering harmless, where the other options are not practicable.
- Technical solutions to control **particulate** emissions to the air are filtration, separation and wet scrubbers.
- Technical solutions to control **gaseous** emissions to the air are adsorption and water walls.

Controlling Air Pollution

Industrial air-pollution control encompasses the design, process engineering and abatement techniques necessary to eliminate, reduce or render harmless the emission of contaminants into the atmosphere. The most cost-effective and efficient methods are those incorporated into the process design to reduce the total mass of contaminants in the waste stream. The engineering devices should be supplemented by management techniques, i.e. procedures, information, instruction and training.

Air pollution needs to be eliminated

Control Hierarchy

The control hierarchy describes a system of controls of different effectiveness. For instance, if a pollutant can be **eliminated** then there is no need to have procedures in place to minimise it, or render it harmless before exhausting to atmosphere. If it can be **minimised** then there is less to deal with. If neither of these are possible, then we are left with the **render harmless** option. One problem with this is that the harmful substance is still in use and if it escapes through failures in process or equipment it can still cause harm.

> ### TOPIC FOCUS
> **Control Hierarchy**
>
> **Eliminate**:
>
> - Replace solvent-based chemicals with water-based chemicals, e.g. paints.
> - Replace chemical process with mechanical process. Mechanically-generated emissions are generally easier to collect than those produced through a chemical process.
> - Replace halogenated products with non-halogenated products. When CFCs were banned, they were replaced with products such as propane and isobutene as alternative propellants (or use pump-action sprays, which remove the need for any propellant).
>
> ### TOPIC FOCUS
> **Minimise**:
>
> - This has been achieved in the motor industry through the use of improved technology, such as engine management systems and fuel injection. Modern cars do significantly more miles/kilometres per gallon/litre of fuel than older cars, when we compare similar engine sizes. They are also more powerful, so both fuel economy and performance have improved, yet emissions are reduced.
>
> **Render harmless**:
>
> - The techniques described in the **Examples of Technology** subsection of this element either minimise pollutants, or render them harmless before they are emitted to atmosphere.

Examples of Technology

There are many techniques available to control pollution to the atmosphere and we discuss some of the main ones below. The choice of technique will depend on a number of variables, such as the:

- type and volume of pollutant to be controlled; and
- environment in which the process takes place.

Filtration

Fabric filters remove dust from a gas stream by passing it through a fabric. The fabric must allow air to pass through it and remove the dust particles from the air. The layer of dust which accumulates on the fabric surface is called the filter cake.

Fabric filters are generally more efficient at removing smaller particles from air streams than cyclones (see later). Consequently, cyclones are often used as first-stage air-cleaning devices to remove the larger particles from the air stream before it is passed into a fabric-filter unit.

Fabric-Filter Types

Fabric filters are normally designed with the fabric forming cylinders or bags. Usually, there are several filter bags or filter elements grouped together in an enclosure; the whole cleaning device is called a 'bag house' or 'bag filter plant'. Types of bag filter plant are differentiated by the mechanism used to remove the filter cake from the surface of the bag.

There are three commonly used mechanisms:

- **Mechanically Shaken**

    In the early 1890s, bag-shaped filters were used and these were shaken by hand to remove the filter cake. Modern bag-filter plants employ mechanical shaking devices to vibrate the bag at frequencies between 10 and 100 cycles per second, for a few minutes. Generally, the bag is open at the bottom and closed at the top. The dust-laden air enters the bag at the bottom and passes up and through the bag to leave the filter plant through vents at the top. The filter cake therefore accumulates on the inner surface of the bag.

    The cleaning cycle is operated at regular intervals to remove the filter cake before the airflow through the bag is stopped, and a slight reverse airflow is sometimes introduced to aid cleaning. The bags are shaken and the released dust is collected in hoppers at the base of the plant.

    The mechanical shaking of the bags induces friction and stresses the fabric, so the material of the filters must be chosen to tolerate this.

- **Reverse Airflow**

    This cleaning technique involves passing cleaned air through the bags in the opposite direction to the normal operating direction. In high-temperature operations the cleaned air is re-circulated rather than using colder ambient air, which reduces the thermal stresses in the plant and prevents condensation.

    During the cleaning cycle, the normal airflow is diverted and a reverse air current applied to the outside of the bag. This change in pressure initially causes the bag to deform and the filter cake is dislodged and falls into a hopper. This method of cleaning involves less mechanical stress to the bags and so the strength of the fabric material is not so crucial.

- **Pulse-Jet Systems**

    Pulse-jet bag-filter plants employ jets of compressed air to remove the filter cake. In these plants, the bag-filter elements are closed at the bottom and open at the top. The dust-laden air passes from the outside of the bag to the inside and up to vents at the top of the plant. The filter cake forms on the outside of the bag. To prevent the bags collapsing in normal operation, they are supported on the inside by metal rings or cages.

    During the cleaning cycle, the airflow to the bags is redirected and air, from compressed air nozzles at the open tops of the bags, is directed into the bags. This positive pressure slightly inflates the bags and the deformation and outward flow of air dislodges the filter cake. The dislodged dust falls into a hopper and is removed from the plant.

5.3 Control Measures to Reduce Emissions

Bag Filter Efficiency

The method of measuring efficiencies involves measuring the particle concentrations in different size ranges and expressing efficiency as the percentage of mass concentration retained by the plant in each size range.

Specific characteristics are important in designing plants to deal with specific situations. The parameters include the gas-to-cloth ratio for particular materials. This is the measure of gas flow through a unit area of material. However, this measure considers only the material and not the filter cake. There are various theoretical equations for pressure drop across a porous bed and they are applied to material and filter-cake combinations to determine the appropriate fan sizes and cleaning cycle frequencies.

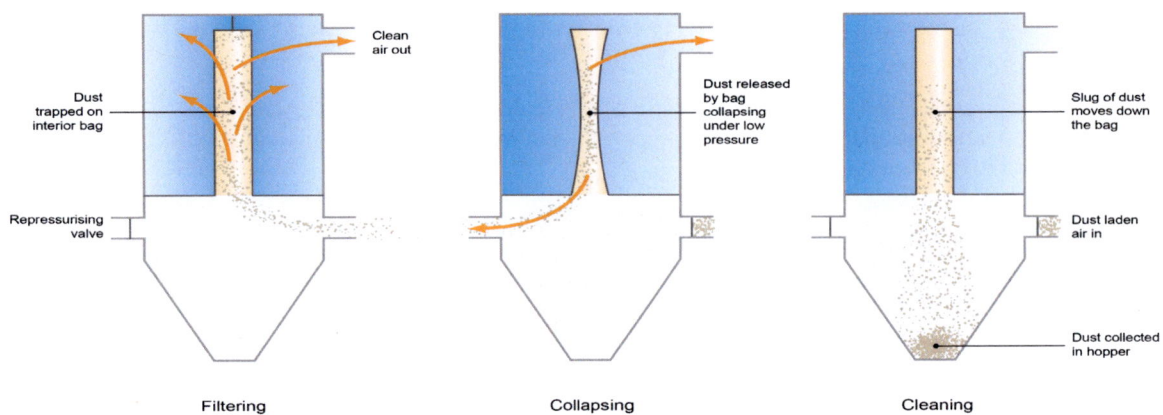

Bag filter reverse-air cleaning

Local Exhaust Ventilation (LEV)

The main object of LEV is to extract the flow of air away from a work process that uses hazardous airborne substances. The air is cleaned, often with a bag filter, before exhausting it to the outside atmosphere.

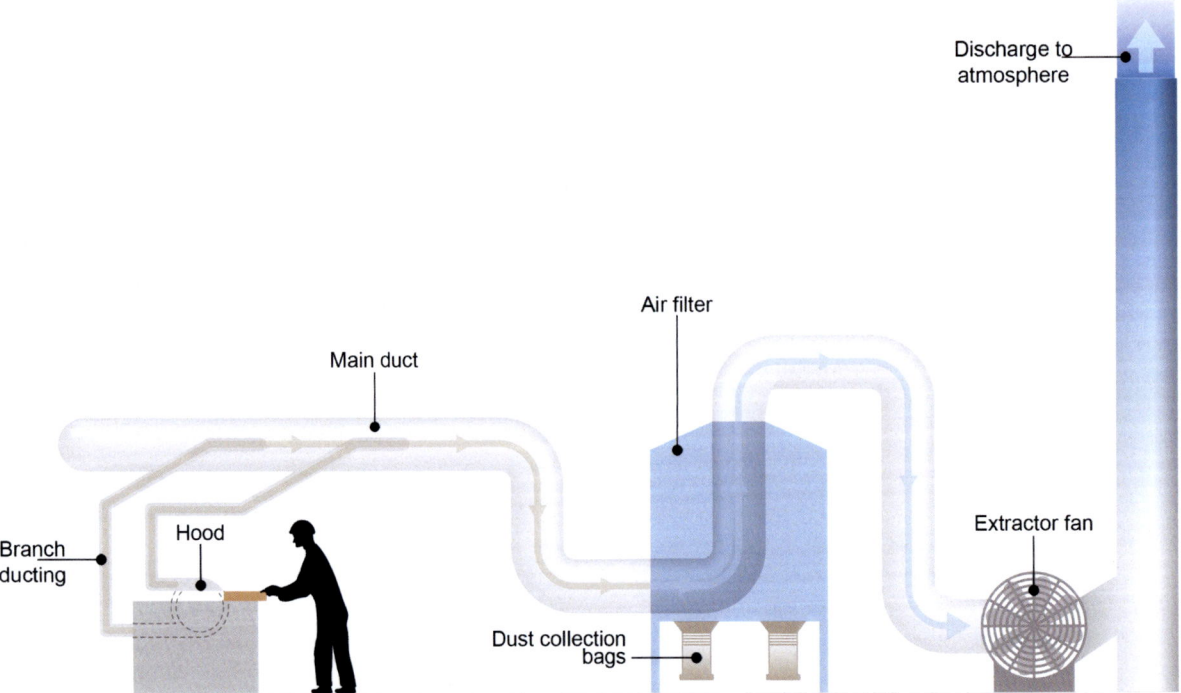

A typical LEV system extracting sawdust from a bench-mounted circular saw

Separation Technology

Gravity Separators

These devices use the force of gravity as the primary method of separation. A settling chamber reduces airflow speed as much as possible for as long as possible. Grit and large dust particles collect in hoppers beneath the chamber. This may be used as a first stage of more thorough cleaning, e.g. to settle fly ash in a power station or Municipal Solid Waste (MSW) incinerator.

Cyclones operate by causing the airflow to change direction rapidly into a spiral, throwing the particles out of the air stream toward the walls of the device. The particles then fall down to the bottom of the device for collection. Cyclones are most efficient for large dense particulates; smaller, less dense particulates may be carried on through the cyclone.

Cyclones are used primarily for the following functions:

- Product recovery, e.g. wood dust.
- First-stage air-stream cleaning.
- Droplet removal.

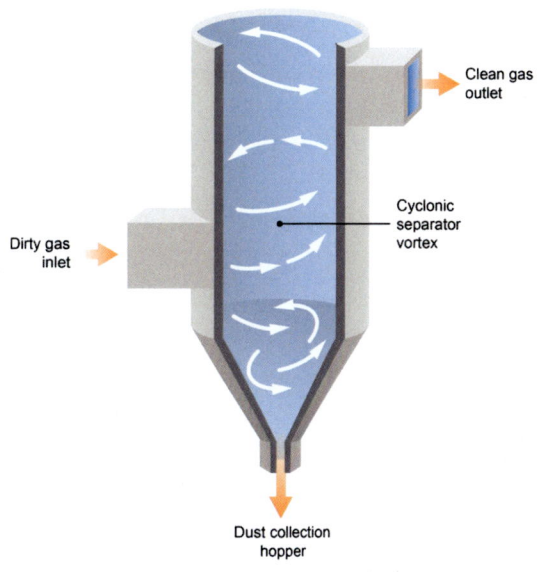

Cyclone schematic

Single cyclones have no moving parts, so the running costs and maintenance requirements are low. However, their efficiencies are much lower than those of fabric filters or electrostatic precipitators (see below) and, generally, they are not suitable for achieving current air-emission standards.

Cyclones may be arranged in groups and operated in parallel.

Electrostatic or Magnetic Separators

- **Principles**

    An electrostatic precipitator (ESP) is a particulate and droplet control device which uses electrical forces to remove particles from a dust-laden air stream. An area of ionised air molecules is established, usually around a wire, by maintaining the wire at a very high voltage, typically 20,000 to 100,000 volts. This region of ionised air molecules is called a corona. As dust particles flow through the corona, they collect the ions, and then the dust particles themselves become charged. Small particles, around one micron, may collect tens of thousands of ions. A plate, called the collector plate, is maintained at the opposite electrical polarity to the wire and the particles, so that the charged particles migrate toward the plate.

> **DEFINITION**
>
> **IONISED MOLECULE**
>
> Physically converted into an ion by adding or removing a charged particle such as an electron.
>
> For example, Sodium (Na) is a neutral atom (it is neither negatively nor positively charged). When it combines with Chlorine (Cl) it loses an electron and becomes positively charged (Na⁺). The Chlorine gains the electron and becomes negatively charged (Cl⁻). They are ionised and combine together to form NaCl (common salt).

5.3 Control Measures to Reduce Emissions

Electrostatic precipitators are normally arranged with a series of wires between rows of plates so that as the particles pass each wire, they collect more of a charge and drift progressively towards the plates. The removal of dust from the plates is often achieved by rapping the top of the plates mechanically, using a hammer or piston. The released dust then drops or slides down the plate into a hopper. During this process, approximately 10% of the dust may re-enter the air stream. Most of this dust is recaptured, but dust released at the outlet of the device will escape into the exhaust air stream.

The dust deposited on the plates is not a solid cake, as in a bag-filter plant, but a fragile deposit. Therefore, there may be re-entry of the dust by the airflow over the plates. To prevent this, baffles are often included to reduce airflow over the plate surface.

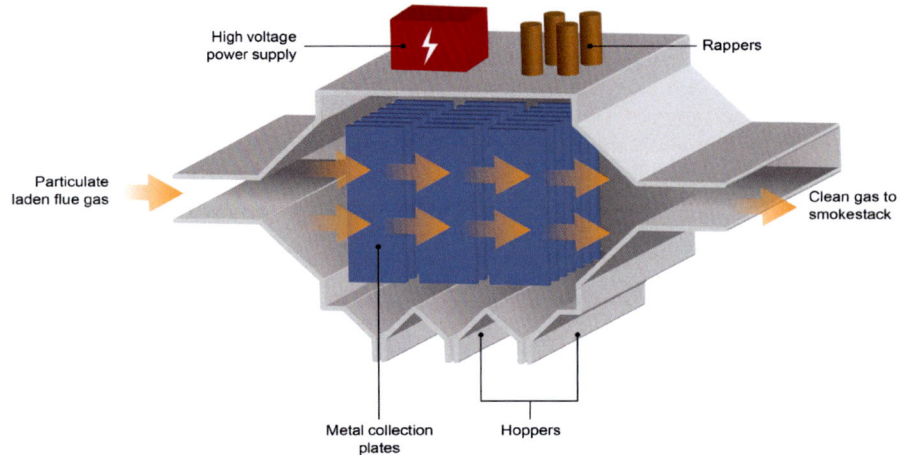

Electrostatic precipitator (plate-wire)

> **TOPIC FOCUS**
>
> **Types of Electrostatic Precipitator**
>
> There are four main types of precipitator:
>
> - **Plate-Wire Precipitators**
>   - As described.
>   - The most common type.
>   - Used in a wide variety of industrial applications, e.g. coal-fired boilers, cement kilns, solid-waste incinerators, paper-mill recovery boilers, petroleum refining and catalytic cracking units, sinter plants, basic oxygen furnaces, open-hearth furnaces, electric-arc furnaces, coke-oven batteries and glass furnaces.
>
> - **Flat Plate Precipitators**
>   - Used for smaller applications.
>   - Use a central plate rather than a wire.
>   - Since a corona cannot be generated on flat plates, needle-like electrodes are located on the leading and trailing edges of the central plates.
>   - Flat plate ESPs have applications for small (less than one micron) particles with high resistivities.
>
> (Continued)

Control Measures to Reduce Emissions — 5.3

> **TOPIC FOCUS**
>
> - **Tubular Precipitators**
>   - Early ESPs were tubular, with the discharge wire running up the centre of the tube. To accommodate higher airflows, the tubes were often arranged in bundles.
>   - The tubes may be formed as a circular, square or hexagonal honeycomb and can be tightly sealed to prevent leaks of material.
>   - Most often used in sulphuric-acid plants, coke ovens, and iron and steel plants.
>   - Often also used to recover valuable materials, or to control the release of hazardous material.
> - **Water-Irrigated Precipitators**
>   - May be of any of the design types discussed above, but with walls washed with water rather than the dry dust rapped from the surface.
>   - Water flow may be continuous or intermittent, with the sludge collected in a sump below the plates.
>   - This method generates slurry, which is more difficult and expensive to dispose of than a dry dust deposit.

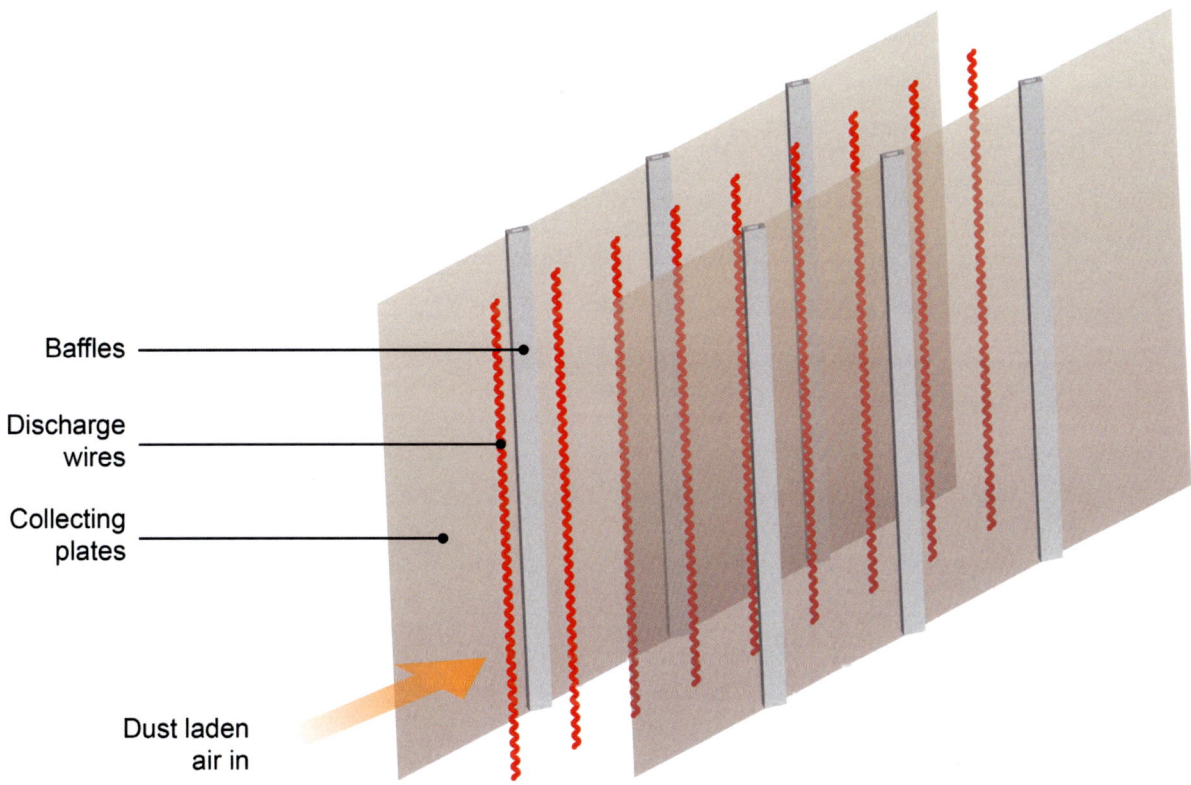

Flat plate electrostatic precipitator

- **Typical Applications**

  Electrostatic precipitators are often used as the final stages in an air-cleaning system. Where there are high dust loadings with large particles, a cyclone is often used as a first-stage cleaning device to remove the coarse, or large particles from the air stream.

5.3 Control Measures to Reduce Emissions

Wet Scrubbers

Principles

Wet scrubbing techniques are used to remove particulates and gases from waste gas streams.

Wet scrubbing techniques are normally employed where the:

- Contaminant cannot be removed easily in a dry form.
- Waste gas stream contains both particulates and soluble gases.
- Particulates to be removed are soluble or wettable; they would adhere to the inner surfaces of a cyclone or bag-filter plant and clog it.
- Contaminant will undergo some subsequent wet process, such as sedimentation, wet separation, or neutralisation.
- Pollution control system must be compact.
- Particulates may ignite, or explode if collected in a dry form.

Wet scrubbing is used to:

- Control sticky emissions, which may block filter-type collectors.
- Handle waste gas streams containing both particulates and gases.
- Recover soluble dusts and powders.
- Remove metallic dusts, such as aluminium, which may explode if handled dry.

The principle of all wet scrubbers is that water droplets are generated within the device and particles are captured within the droplets. The droplets are then removed from the air stream, which is now clean. The droplets are collected as contaminated water and transported out of the device for treatment or disposal. To aid the removal of contaminants, the water droplets may be acidified (made more acidic), or basified (made more alkaline).

The methods for bringing the water droplets in contact with the dust-laden air include:

- Injecting water directly into the air stream and mechanically shearing the water into droplets.
- Spraying the water into the gas stream.
- Injecting water onto a spinning disc or fan.

Different scrubber designs use different techniques, or combinations of techniques.

TOPIC FOCUS

There are five basic types of **scrubber design**:

- **Venturi Scrubbers**

  These devices create atomised droplets by injecting water into the gas stream before accelerating the water through a high-velocity zone called a venturi throat. The water and the gas stream are then released into a low-pressure area called the diverging section. The turbulence in the venturi throat breaks the water into tiny droplets and particle capture occurs toward the end of the venturi throat and at the beginning of the diverging section.

- **Mechanically-Aided Scrubbers**

  These devices use spinning discs or fans to generate water droplets.

- **Pump-Aided Scrubbers**

  These devices spray the water as droplets into the gas stream.

- **Wetted Filter Scrubbers**

  These devices use a combination of water spray and a filtration element. Particles are captured by water droplets, as described previously. However, particles may also impact temporarily on the elements of the filter to be washed off by a film of water.

- **Tray or Sieve Scrubbers**

  Tray or sieve-type wet scrubbers have small holes in trays that accelerate the gas stream. Water is piped onto the trays to form a shallow layer of water. The airflow through the holes creates a froth, which assists in capturing particles.

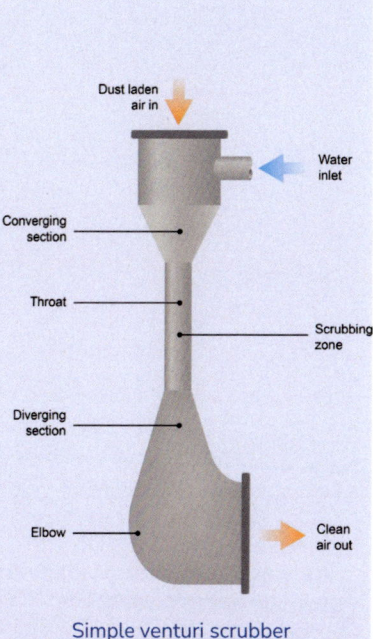

Simple venturi scrubber

Droplet Removal

Many scrubbers use cyclonic separators or cyclones to remove droplets. Others use **chevron** droplet eliminators for either vertical or horizontal gas flow. Shaped like curved and parallel blades, the chevron introduces a surface against which droplets impact and accumulate as water and then drain off. The solids that accumulate on the surface are periodically washed off using water sprays.

For finer droplets, mist eliminators comprising a fine metal mesh are often used. A layer of wire mesh is introduced in the final duct and the mist accumulates on it and drops off. The mesh mist eliminators are also spray-washed periodically to remove any particulate build-up.

Adsorption

The process of adsorption involves the retention of a gas or vapour molecule on the surface of a particle or droplet. The phenomenon is essentially a surface reaction, as opposed to absorption, which involves the complete encapsulation of a molecule, which is then dissolved in a liquid droplet. Some solids with many pores and crevices present extremely large surface areas to gases and so are the most appropriate adsorbents. These include activated carbon, activated alumina and silica gel.

5.3 Control Measures to Reduce Emissions

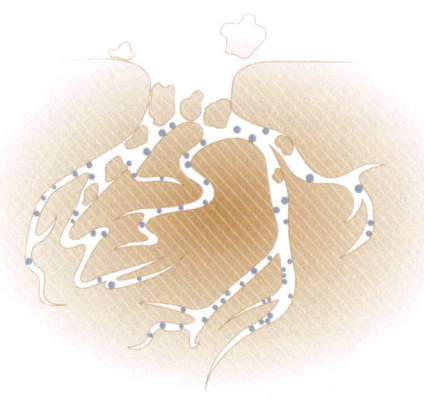

Adsorption on a solid with many pores

Adsorbents are selected depending on the type and quantity of contaminant you wish to remove:

- **Activated Carbon**

    Activated carbon is charcoal that has been heated in the absence of air. At one time, wood was heated to produce charcoal, but later developments include the use of coal, coconut shells, peat and other substances.

    After heating, the carbon is activated to remove the volatile components. In the case of coal, high-temperature steam is used. However, zinc chloride, magnesium chloride, calcium chloride and phosphoric acid have also been used as activating agents.

- **Activated Alumina**

    Activated alumina and hydrated aluminium oxide is produced by special heat treatment of aluminium ore or bauxite. Activated alumina is mainly used for drying gases under pressure, as it has an affinity for water.

- **Silica Gel**

    Silica gel is an amorphous form of silica, derived from the interaction of sodium silicate and sulphuric acid. It is often used as an adsorbent where activated carbon is not appropriate. Like alumina, silica gel has an affinity for water.

Operational Mechanisms

Adsorption systems are designed either to:

- remove pollutant gases and vapours from air streams, to prevent the emission of those pollutants to atmosphere; or
- collect those vapours to return them to the process.

In either case, there are four phases in the process:

1. Contact between the polluted air stream and the adsorbent under conditions that allow adsorption of the pollutant.
2. Removal of the cleaned air stream from the adsorbent.
3. Regeneration of the adsorbent to recover the pollutants and re-use the adsorbent.
4. Re-use or disposal of the pollutant.

The adsorbent most often used is activated carbon.

- **Static-Bed System**

  In simple systems, granulated activated carbon is held in a vertical column and solvent-laden air is passed down through the column. The solvent is progressively adsorbed on the carbon and the cleaned air passes out of the column to the atmosphere. After a predetermined period, set to ensure that the carbon is not completely saturated with solvent, the airflow through the column is shut off and the carbon is regenerated by lowering the gas pressure, or by increasing the temperature. This desorbs (releases) the contaminants, which can be recovered.

  After the solvent is steam-stripped, the carbon beds are hot and saturated with water. The beds are normally opened and air-dried, allowing the water to evaporate to atmosphere. Multiple systems are common where two or more columns are used. This allows some columns to be in the adsorption part of the cycle, while others are in the regeneration part of the cycle.

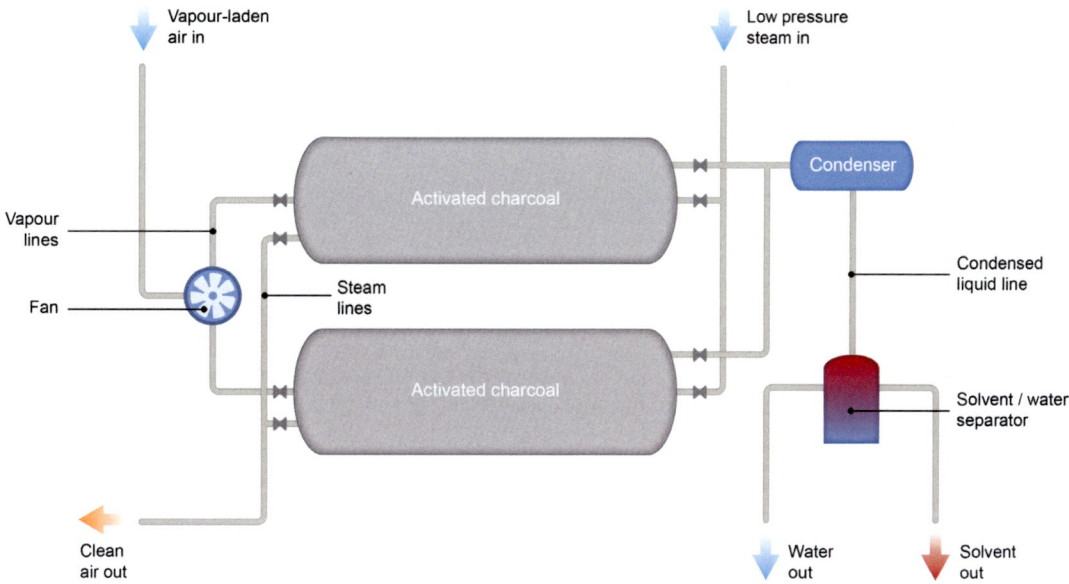

Simple activated carbon solvent recovery system

- **Rotary-Bed Systems**

  In order to ensure more efficient use of the carbon bed, continuous rotary-bed systems have been developed. These consist of a rotating drum containing activated carbon. The drum has a hollow central core along the axis of rotation, and the space between the inner and outer walls of the drum is divided into radial sections. Vapour-laden air enters the drum in one section, at one end of the drum. It then travels along the length of the section and the vapour is adsorbed in the carbon. The cleaned air leaves through the central core from the far end of the drum.

  Once that section is saturated with vapour, the drum rotates to the next section. At another vapour-saturated section of the drum, steam is pumped up a pipe in the central core, to enter the section at the far end of the drum. The steam passes through the vapour-saturated carbon to exit as a steam and solvent mixture at the front end of the drum. Therefore, there is always adsorption and regeneration within the drum.

- **Process Controls**

  Before a carbon-bed adsorption system is considered or designed, careful consideration should be given to modifying existing processes and procedures to reduce the quantities of VOCs in the exhaust air streams:

  - Consider whether the use of the solvent is necessary, or whether a water-based system or detergent degreasing system could be used.
  - Consider the substitution of the solvent for a lower volatility solvent, or a less toxic solvent, or one with a lower environmental impact.
  - Minimise the ventilation rates and volumes in the process to reduce evaporation rates.

5.3 Control Measures to Reduce Emissions

- Establish working procedures for effective use of the system and train workers to comply with them.
- Provide well-designed LEV systems with hoods and tank enclosures.
- Cover tanks when not in use.
- Perform solvent spraying in a vapour zone. Do not use compressed-air drying techniques.
- Do not direct ventilation fans onto solvent baths, containers, or uncontrolled drying areas.

- **Maintenance and Operation**

The surface area of carbon granules must be protected against dirt and other particulates entering the bed. It is common to have a fabric filter, or bag filter as a primary air-cleaning device located upstream of the carbon bed. Some solvents entering the adsorption bed will chemically react and progressively reduce the working surface area of the carbon. Such substances must not be allowed to enter the bed.

Many LEV systems have been introduced to satisfy occupational hygiene requirements. Poor hood and enclosure design and leaking ducts have often been compensated for by higher ventilation velocities and volumes, which are not consistent with efficient final air-cleaning characteristics. Careful consideration should be given to improved design characteristics, which deliver lower ventilation velocities and volumes. This will lead to:

- Lower space-heating energy requirements.
- More efficient final air cleaning.
- Lower atmospheric emissions.
- Lower fan and system maintenance costs.

Water Walls

Water can be used to suppress dust and prevent it escaping from a confined area. This technique is commonly used in construction and demolition, where large amounts of dust can be generated and therefore need to be controlled. There are two common methods:

- **Rain guns** - used to spray water over stockpiles or buildings being demolished and keep material damp but not wet. This reduces the amount of dust created by the break-up of the building material. The guns can be either static or mounted on vehicles such as loading shovels so they suppress dust as material is moved around a site.
- **Perimeter systems** - create a curtain of fine water particles that can cover a large area where activities may create dust.

With both systems the design of the nozzles is critical to avoid using excessive volumes of water and creating run-off that then needs to be controlled. If designed and installed correctly there should be little, if any, run-off as the particle size of the water droplets is such that they encourage dust particles to bind together and therefore drop out of suspension in the atmosphere.

Combustion Devices

Thermal oxidisers is a mechanism for air pollution control that causes the destruction of hazardous gases at elevated temperatures prior to release into the atmosphere. They are often used to render harmless VOCs from an airstream by thermal combustion to produce carbon dioxide and water vapour. Organic compounds will ignite at around 620°C; a catalyst may be used to lower this temperature.

Maintenance of Equipment

Any equipment that is used will need to be subject to planned preventative maintenance. This will ensure that it continues to carry out the abatement of air pollutants to the initially planned level. It is important, therefore, that funding for maintenance is provided to ensure the devices' continuing effectiveness.

Such funding may be difficult to obtain in developing countries. Poor maintenance of pollution abatement and other equipment is an important cause of process failure and pollution incidents in developing economies, often because of a lack of proper funding. It should also be recognised that certain environments present significant logistical or practical challenges in undertaking regular planned maintenance - for example, on offshore oil and gas installations, or facilities located in Arctic or desert environments.

Control Measures to Reduce Emissions — 5.3

TOPIC FOCUS

Summary of abatement technologies

Abatement Technology	Uses	Mode of Operation
Fabric Filters	Removal of particulates - efficient for small particles.	Dust-laden air is passed through a filter before being extracted to the outside. The dust is caught on the filter and removed by mechanical means, reverse airflow, or pulse jet systems.
Gravity Separators	Removal of particulates: • Only used for larger particles. • Often used before a fabric filter.	Dust-laden air is passed through a settling chamber at a lower air speed. Grit and dust drops and is collected in a hopper below the chamber.
Cyclones	Removal of particulates: • Only used for larger particles. • Often used before a fabric filter.	The dust-laden air is rapidly forced into a spiral, causing particles to fall to the bottom of the device for collection.
Electrostatic Precipitators	Removal of particulates - efficient for small particles.	Dust particles are electrically charged and attracted to a plate of opposite charge. The collected dust on the plate can then be removed mechanically.
Wet Scrubbers	Removal of particles and gases.	The particles or gases are brought in contact with water droplets (or other liquids, if collecting some gases). The particles are captured in the droplets and removed as a sludge.
Adsorption	Particularly useful for removal of VOCs.	Activated carbon is the most commonly used adsorbent. The contaminated dust is passed over the activated carbon and 'captured' within the surface pores (but doesn't chemically react). The application of heat will cause the VOCs to be released and captured. The activated carbon can then be re-activated and used again.
Water Walls	Usually used for the suppression of dust.	Water may be sprayed over stockpiles of dusty materials to prevent them becoming airborne, or a curtain of fine water may be placed around an area where dusty work is being undertaken.

5.3 Control Measures to Reduce Emissions

STUDY QUESTIONS

3. List the three options that make up the control hierarchy and give an example of each.
4. What is wet scrubbing and when might it be used?
5. What is the objective of local exhaust ventilation?

(Suggested Answers are at the end.)

Control Measures to Reduce Emissions | 5.3

Unit EMC2: Environmental Practical Application

Now that you have studied approximately half of the course, you should be in a position to send your tutor a rough outline of your practical application.

You should have already decided on your chosen area and approached management to ensure that they are happy to co-operate in terms of providing information. You should also have discussed with them any confidentiality issues that may exist. Now you should send your tutor a brief outline of the area you intend to cover and the issues you expect to encounter there.

There are a couple of points to remember before you submit this:

- The area must be sufficiently simple and small to allow you to complete the practical application within three hours (even if this means selecting a small area within the site, such as a warehouse, maintenance depot or single production area, if your site is large).
- The NEBOSH template to be used for the practical application (explained in your guidance for Unit EMC2):
    - Is designed to cover the principal topics contained in the syllabus.
    - Will help you to structure your practical application work.
    - Will help you in completing your outline for your tutor.

You can see from it the kind of issues that NEBOSH expects you to cover in your practical application, so don't forget to look at it. Some sections in the template may not be relevant to your particular site, but your chosen area does need to cover a sufficiently wide range of topics.

You can submit your outline plan for your practical application to your tutor by e-mail.

If you have any queries on the template before you submit your outline, contact your tutor for help.

Summary

This element has dealt with the control of emissions to air. In particular, this element has:

- Outlined how poor air quality can have detrimental effects on human health (e.g. irritation and inflammation of the airways) and can affect the general environment (e.g. acid rain).
- Explained that targets can be set for the maximum acceptable airborne contamination to be achieved within a specified timescale.
- Outlined different sources of air pollution, including fossil fuels, industrial processes and transport.
- Described common pollutants of the atmosphere, including:
    - Sulphur compounds (e.g. acid rain).
    - Nitrogen compounds (e.g. photochemical smogs).
    - Halogens and their compounds (e.g. ozone depletion).
    - Volatile organic compounds (e.g. formation of ground-level ozone).
- Explained the control hierarchy relating to emissions to air, which consists of:
    - Eliminate, e.g. replace solvent-based paints with water-based ones.
    - Minimise, e.g. improved technology.
    - Render harmless, e.g. use of filtration, etc.
- Outlined how the choice of technique used to control pollution to the atmosphere will depend on a number of variables, such as type and volume of pollutant and the environment in which the process takes place. Available techniques include:
    - Various types of filtration (e.g. bag filters).
    - Separation techniques (e.g. gravity separators).
    - Wet scrubbers.
    - Adsorption.
    - Water walls.

Exam Skills

Question

Scenario
The manufacturing company for which you work is planning to reduce greenhouse gas emissions in its supply chain, following the reduction of direct emissions from the facility.

Task
What are the types and sources of indirect greenhouse gas emissions from the organisation's activities?

(8 marks)

Approaching the Question

Think about the steps you would take to answer the question:

- Read the scenario carefully. With this question you need to develop sources of greenhouse gases that could occur indirectly from a manufacturing company. Note: the question states 'indirect' sources not 'direct'; there are also other greenhouse gases as well as carbon dioxide!
- Now look at the task – prepare notes on:
    - Potential types of greenhouse gases that could be released indirectly.
    - The activities that lead to the release of these greenhouse gases.
- Consider the marks available. In this case there are 8 marks available so you should provide at least 8 sources with their associated emissions.
- Read the scenario and task again to make sure you understand them and have a clear understanding of indirect activities that could cause the release of greenhouse gases.
- Jot down an outline plan - this might include:
    - Off-site energy generation, extracting and processing raw materials, transportation of materials, product use, removal of sinks, product disposal.

Now have a go at the question yourself.

Exam Skills

Example of How the Question Could be Answered

The sources of indirect greenhouse gas emissions from the organisation include those emissions from inputs into the organisation in addition to those associated with the use or disposal of the product. They include:

- Electricity could be generated off-site from the burning of coal or gas, which will release large quantities of CO_2 and other greenhouse gases into the atmosphere.
- Extracting and processing raw materials will also result in emissions of numerous greenhouse gases from the energy required for such processes.
- The materials will also need to be transported to the site. This will result in the burning of petrol and diesel, which will emit large quantities of CO_2 into the air.
- The use of the product manufactured at the site may also cause large quantities of greenhouse gases to be released.
- Greenhouse gas sinks such as forests may also be removed if the organisation uses wood as part of its processes. This will lead to higher levels of CO_2 in the atmosphere.
- A product at the end of its life will be wholly or at least partly disposed of. The most common disposal technique is landfill. Some waste materials when buried will be anaerobically digested, causing the release of methane, which is a potent greenhouse gas.
- Activities of contractors who work on behalf of the organisation could release greenhouse gases through activities such as transportation.
- Staff may commute between their homes and their worksites causing the release of carbon dioxide into the atmosphere.

Element 6

Control of Environmental Noise

Learning Outcomes

- Understand the importance of reducing environmental harm; identify sources of noise; and suggest suitable control measures.

Learning Objectives

Once you've read this element, you'll be able to:

1. Demonstrate awareness of the environmental impacts of noise.

2. Identify sources of environmental harm and suggest suitable control measures for noise.

Contents

Sources and Effects of Environmental Noise	**6-3**
The Characteristics of Noise which Lead to it Being a Nuisance	6-3
The Effects of Noise	6-4
Legal Considerations	6-5
Common Sources of Environmental Noise	6-5
Methods for the Control of Environmental Noise	**6-8**
Monitoring Requirements and Arrangements	6-8
Basic Noise Control Techniques	6-8
Management Controls	6-11
Summary	**6-12**
Exam Skills	**6-13**

Sources and Effects of Environmental Noise

IN THIS SECTION...

- Environmental noise emanates from a number of sources, including industry, road traffic, leisure activities, and domestic neighbouring properties.
- Low-frequency noise is associated with effects such as nausea, headaches and feelings of anxiety.
- Environmental noise may lead to loss of sleep and stress.
- Noise can cause disturbance to local wildlife.

The Characteristics of Noise which Lead to it Being a Nuisance

Noise may be defined as unwanted sound. People exposed to very high levels of noise, e.g. from machinery and equipment in enclosed spaces, can suffer damage to their hearing, and this is a recognised occupational health issue. But noise can also be unwanted because it causes disturbance or annoyance. Most people will have experienced unwanted noise that interferes with life at home or perhaps has prevented a good night's sleep in a hotel room. Noise from a wide variety of sources can travel considerable distances and affect the peace and enjoyment of life. This type of noise is referred to as **environmental noise**, or **noise nuisance**.

Noise - unwanted sound

Noise emanates from a wide range of man-made sources, including:

- Industrial premises.
- Transportation systems.
- Construction sites.
- Agricultural activities.
- Aeroplanes.
- Entertainment and leisure establishments.
- The homes of neighbouring residents.

The perception of noise can be quite subjective. A particular noise may be acceptable to one person, but very annoying to another. The main factors that affect the perception of noise are:

- **Loudness**: loud noise is likely to be more intrusive.
- **Pitch**: low-pitched (or 'low-frequency') sound, e.g. from heavy machinery or the bass from entertainment sound systems, can travel substantial distances. Pulsating low-pitched sound was a particular problem with early wind turbine designs and may also be associated with heavy road traffic.
- **Incidence**: noise that happens only occasionally may be tolerable, but regular noise may be anticipated by people who are affected, leading to a greater sense of annoyance and anxiety. However, types of low-incidence noise such as sirens and the use of explosives have a high nuisance-causing potential.

6.1 Sources and Effects of Environmental Noise

- **Background levels**: a given sound will travel further and be more noticeable in areas that are generally peaceful, such as the countryside, than in a busy city street.
- **Speech**: such as tannoy, loudspeaker systems and public address systems.

Complaints about noise are growing. Many complaints of neighbourhood noise arise as a result of:

- Anti-social behaviour.
- Poor planning controls.
- The juxtaposition of incompatible land uses.
- Specific one-off events, such as clay pigeon shooting, burglar and theft alarms (especially their repeated, intermittent, high-frequency nature), fireworks, explosives, parties or the use of a jack-hammer on a road surface.
- Sirens or other noise interfering with use of tannoy systems for communication.

As we saw in Element 1, unreasonable noise is considered a nuisance.

The Effects of Noise

Noise nuisance can affect the quality of life in a number of ways, by:

- Being intrusive and annoying.
- Resulting in severe lack of sleep.
- Producing headaches and nausea.
- In extreme cases, causing anxiety, stress and depression.

Environmental noise may also affect wildlife. Wild animals may be disturbed and prevented from feeding in certain areas because of man-made noise. Certain animals - notably birds - use complex calls to communicate with other members of the same species, especially during breeding. Man-made noise can interfere with these communication systems.

Typical decibel levels associated with different noise sources

Measurement in dB(A)	Sound
0	The faintest audible sounds
20-30	Quiet library
50-60	Conversation
65-75	Loud radio
90-100	Power drill
140	Jet aircraft taking off 25m away

> **DEFINITION**
>
> **dB(A)**
>
> Noise is measured in decibels (dB). The human ear is, however, more sensitive at the frequencies of 1 kHz to 4 kHz. When the noise is measured, the equipment is adjusted to best represent the way the human ear hears it (dB(A)).

Sources and Effects of Environmental Noise 6.1

Legal Considerations

Environmental noise is often regulated. The nature of such regulation will vary in different jurisdictions; common regulatory mechanisms include:

- Nuisance - regulators will often have powers to issue notices, impose fixed penalties and seize equipment.
- Construction - regulators may have powers to issue a notice to construction sites relating to the noise controls and the duration and manner in which activities are undertaken.
- Strategic action - noise maps and action plans are often to be developed by governments.
- Equipment - there are limits on equipment that creates noise to which manufacturers must adhere, such as equipment intended for outdoor use.
- Planning - environmental noise is sometimes a significant factor in planning decisions and will, where significant, be covered in the legal requirement to undertake a planning environmental impact assessment.

Noise threshold limits are often incorporated in the requirements above and these should be adhered to if relevant.

Common Sources of Environmental Noise

Manufacturing and Related Commercial Activities

Manufacturing activities, especially traditional heavy industries, may generate significant noise from:

- Pressing and forging metal parts (e.g. shipyards and vehicle assembly plants).
- Turbines (e.g. electricity generation plants).
- General machinery (e.g. motors, grinding and planing, air compressors, conveyor systems).
- Extraction systems (e.g. motors, fans).
- Public address systems (e.g. in warehouses and distribution centres).

The main problems arise when domestic premises or institutions, such as schools and hospitals, are located in close proximity to industrial facilities.

Transport Noise

Road traffic, especially heavy trucks, generate considerable noise from:

- Engines.
- Movement over road surfaces, especially from the bodywork of empty vehicles.
- The use of horns.
- Reversing alarms which are often high-pitched and can travel substantial distances.

Large jet aircraft engines are intensely noisy, especially during take-off and landing.

Road traffic is a major source of noise in city centres; even in rural areas, noise from motorways can be detected several miles away, especially at night. Particular problems can be experienced where distribution centres that operate around the clock are located in close proximity to residential areas.

Noise around major airports, especially associated with plans for the expansion of existing airports, has become a significant political issue.

Agricultural Noise

Modern agriculture is highly mechanised. Noise from agricultural activities is often associated with the use of:

- Mobile machinery, e.g. tractors and harvesters, especially when harvesting is undertaken around the clock.
- Bird-scarers that simulate loud gunshots.

6.1 Sources and Effects of Environmental Noise

Any source of noise may be more noticeable in the countryside, because background noise levels are generally lower than in an urban environment.

Construction Noise

Construction work typically involves activities that are potentially noisy, including:

- The use of heavy equipment, such as excavators and cranes.
- Pile driving.
- The use of powered tools, such as drills.
- Trucks and vans delivering materials and removing debris.

Construction work often takes place near town centres and in residential areas, and is a major source of environmental noise complaints. A balance often needs to be struck between the needs of the developer in completing the construction project and the avoidance of unacceptable disturbance to local residents.

Construction and demolition can cause substantial environmental noise

Quarrying and Mining

The extraction of minerals from the ground also involves activities that are potentially very noisy, including:

- Operation of excavation machinery.
- The use of explosives.
- Operation of rock-crushing equipment.
- Operation of conveyor systems.
- Movements of heavy vehicles on and off site.

Quarrying and mining can be an important source of noise nuisance in rural areas.

> **MORE...**
>
> For more information on noise nuisance go to:
>
> https://noisenuisance.org
>
> www.nidirect.gov.uk/noise-nuisance-and-neighbours
>
> www.gov.uk/report-noise-pollution-to-council

Noise from Pubs and Clubs

Noise nuisance may be caused by the operation of sound systems at clubs and open-air festivals or by people arriving and leaving. Noise may equally cause problems to residents in proximity to sporting events, such as motor car racing circuits, or open-air music concerts which are often located in quiet, rural areas.

Limits may be set in relation to the:

- Number of times such events are permitted.
- Distance from noise-sensitive premises.
- Amplitude of music from loudspeakers.
- Time and duration of such events.

Sources and Effects of Environmental Noise 6.1

Neighbour Noise

Environmental noise is most likely to disturb us in our own homes. Noise nuisance is often created by the activities of other residents in the neighbourhood. Sources of noise that commonly cause annoyance include:

- Loud music from radios and other sound systems.
- Televisions.
- Use of equipment such as power drills, lawn mowers and strimmers.
- Dogs barking persistently or late at night.

Noisy neighbours is a problem that is increasingly being dealt with by local authorities and can be exacerbated by:

- People living closer together.
- Modern construction techniques providing limited sound insulation.
- Ever decreasing levels of social tolerance.

Intruder and Vehicle Alarms

Noise from all intruder alarms can be a nuisance.

The standards for intruder alarms often relate to how the alarms are activated, to ensure few false alarms and the silencing of any siren after a period of time (usually 20 minutes).

Wind Farms

The construction phase of a wind farm can lead to significantly high noise levels. These will result from the construction activities considered earlier.

Older designs of wind farms sometimes created a low-pitched pulsating noise that could travel significant distances. However, noise from the operation of modern wind turbines is usually minimal.

Rural Noise

As we have briefly mentioned, noise nuisance is not restricted to busy urban environments. Noise nuisance can be a significant problem for the residents of rural areas. Whilst overall noise levels are likely to be higher in urban environments, individual noises may be more noticeable in the countryside. Sources of rural noise nuisance include:

- Mobile farm machinery, especially during harvest time.
- The operation of bird-scarers.
- Noise from motorways and railway lines that traverse the countryside.
- Mining and quarrying activities that are typically located in rural areas.
- Animals with loud and persistent calls, such as dogs, cockerels and peacocks.
- Open-air festivals and motor racing circuits.

> **STUDY QUESTION**
>
> 1. List five different sources of potential noise nuisance.
>
> (Suggested Answer is at the end.)

6.2 Methods for the Control of Environmental Noise

Methods for the Control of Environmental Noise

IN THIS SECTION...
- Noise control techniques include physical controls such as: isolation, absorption, insulation, damping, and silencing.
- Management controls may also be implemented, covering hours of working, vehicle routes, etc.

Monitoring Requirements and Arrangements

Requirements for monitoring of noise are often incorporated into a formal environmental management system. There may be noise monitoring requirements stated in legal controls such as environmental permits.

In order to assess the impact of industrial and commercial sound, the British Standards Institution has published BS 4142:2014+A1:2019 *Methods for rating and assessing industrial and commercial sound*. This standard sets out a method for assessing industrial and commercial sound level against the existing background sound level, to determine the likelihood of adverse impacts.

Basic Noise Control Techniques

Wherever noise is a problem, the order of priority for dealing with it is:

- **Noise reduction at source** - e.g. by elimination or substitution of the process or equipment producing the noise.
- **Attenuation in transmission** - by engineering controls that limit the amount of noise transmitted.
- **Protection of the receiver** - e.g. through the use of double-glazing and the design of houses with living rooms and bedrooms away from roads.

Elimination, Substitution and Maintenance

Noise can often be eliminated or reduced by replacing noisy equipment or processes with alternative, quieter equipment or processes. For example:

- Diesel/petrol engines replaced by electric motors.
- Pneumatic tools replaced by electric tools.
- Solid wheels replaced by pneumatic rubber tyres.
- Metal chutes, buckets or boxes replaced by rubber or plastic ones.

Many machines are noisy because of worn parts, poor maintenance, inadequate lubrication, or because they are 'out of balance'. Planned maintenance, replacement of worn parts and regular oiling will reduce noise and increase efficiency.

Rather than replace a complete machine or process, it may be possible to carry out a simple modification, e.g. plastic or rubber-coated rollers and guides on a conveyor belt may be used for handling glass or metal components.

Pneumatic tools could be replaced by electric tools

Methods for the Control of Environmental Noise | 6.2

Isolation

In many cases, the best method of noise control is to enclose the noise source. Machinery enclosures should:

- Have a heavy, noise-reflecting outer skin.
- Have a noise-absorbent lining, such as mineral fibre.
- Be mounted so that they do not transmit noise and vibrations to the floor.
- Be airtight - the smallest gap allows sound to escape and reduces the attenuation of the noise inside the enclosure. This is a particular problem with, for example, woodwork machines, such as saws and planes, where timber is fed in at one end and comes out at the other. Such equipment can, however, be fitted with noise-reducing feed and delivery tunnels which should be lined with noise absorption materials and fitted with windows to allow clear viewing, and with adequate lighting.

Absorption and Insulation

Machines are often situated in large, acoustically reverberant areas which reflect sound and build up noise levels within the room. Noise levels in adjacent rooms can be reduced significantly by using sound-absorbing materials on walls and other large surfaces. The absorptive surfaces reduce the reverberant component of the overall sound and, consequently, the level of noise in general.

As well as possessing absorbent properties, noise screens or enclosures and havens must be acoustically insulating. This means that they must transmit very little noise and therefore tend to be heavy. The superficial density of the barrier must be high.

Environmental Noise Barriers

Screening the noise source is a common way of preventing noise spread. Suitable barriers could be:

- High walls or fences.
- Purpose-built earth berms or bunds.
- Other buildings in the vicinity of the noise source.

> **DEFINITION**
>
> **BERM**
>
> A raised barrier usually made of compacted soil.

Although noise in some ways resembles the behaviour of light, in others it does not, and noise can be refracted round obstacles, literally travelling round corners. Therefore, hiding a piece of noisy equipment from view will not stop the noise.

Placing a barrier, such as a wall or fence, will only have a limited effect. This is particularly the case with low-frequency sounds. High-pitched noises tend to behave more like light and can be screened more effectively.

Screens may reduce noise from a small piece of equipment by preventing the noise escaping in a particular direction. Screens should be placed near to the source; the greater the angle, the better the noise reduction. Such a screen could reduce noise by 5 to 10 dB.

6.2 Methods for the Control of Environmental Noise

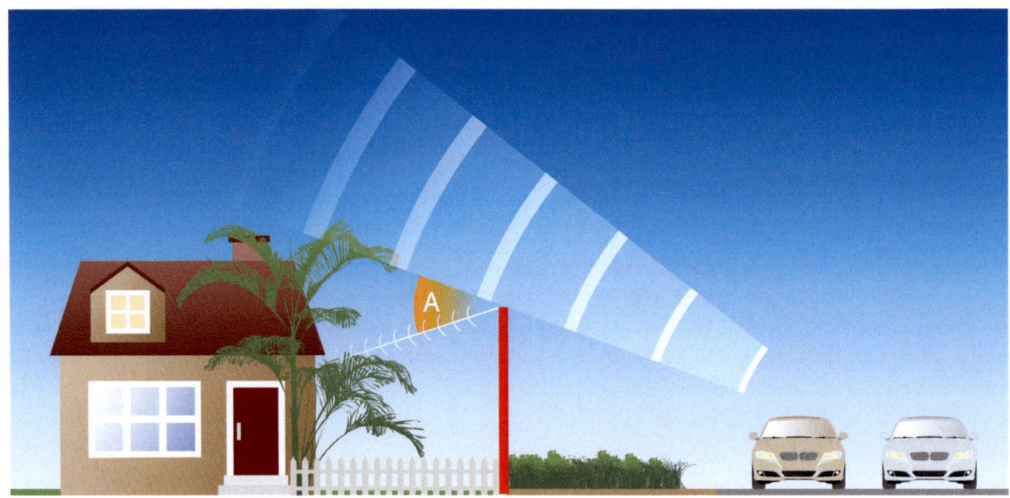

An acoustic barrier (which should be close to the source and as high as possible, to increase angle 'A' for more noise reduction)

Damping

Vibration is one of the main causes of noise. These vibrations can be transmitted from the source, via a rigid connection, to a variety of sites, such as the panels of a machine, floors, walls and tables. These large surfaces act as sounding boards and increase the level of noise.

The simple expedient of isolating the machine on anti-vibration dampers or rubber mountings may reduce noise levels considerably, e.g. putting rubber feet around the legs of machines. Other damping techniques include construction methods using bolts rather than welds and surface coatings or bonding applied to sheet metal.

Silencing

Certain types of equipment involving the intake or discharge of air or other gases may be fitted with acoustic silencers, similar to the way in which gunshot sounds or the noise from car exhausts may be prevented. These work by absorbing the sound pressure generated by the process at its source.

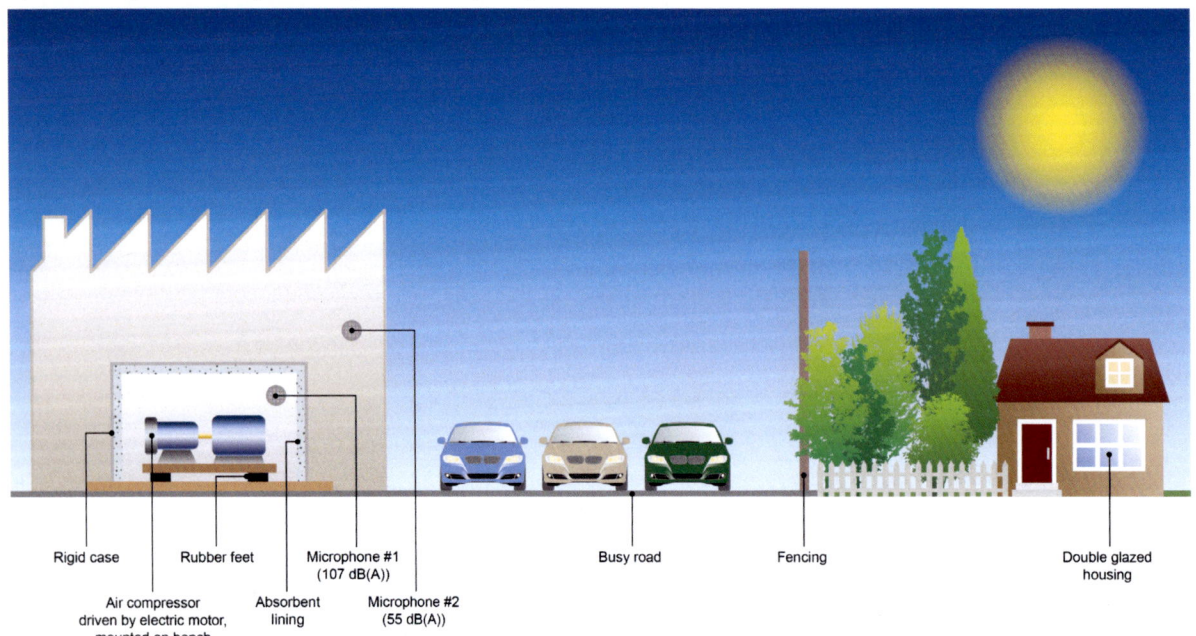

Some methods of controlling noise

Management Controls

Management controls are potentially more effective than the physical controls described above as, if implemented effectively, they will prevent the production of noise in the first instance and so move the controls up the hierarchy of control discussed above.

Examples of management controls are:

- **Maintenance regimes** - as stated earlier, equipment that could cause significant noise release should be subject to planned preventative maintenance.
- **Control of working hours** - usually to reasonable daytime hours. Most people are out and busy during the day and, because there is more general activity taking place, individual noise sources are less likely to cause a problem.
- **Controlling the use of radios** (both music and two-way radios) - radios used for communication and entertainment can cause a nuisance to others nearby. Controlling the number of radios on a site and the volume they are played at is essential. When using two-way radios, the use of earpieces has advantages; they:
    - Prevent others hearing conversations and therefore any potential to cause a nuisance.
    - Improve the ability of the user to hear what is being said over the radio.
- **Public address systems** - should be designed so that sound is directed where it needs to be heard and not beyond the boundaries. This may mean using more, smaller speakers and being able to reduce the volume at night when background noise levels are generally lower.
- **Vehicle routes** - vehicles entering and leaving premises, especially large goods vehicles using air brakes and often air-assisted gear changes, can create significant noise levels. Proper routeing of these vehicles, together with signage indicating any prohibited areas or routes, can reduce the nuisance caused. Driver training can also have a significant positive effect.
- **Loading doors and shutters** - ensuring these are kept closed when not in use, especially during the night, can significantly reduce the noise levels and the potential for nuisance.

STUDY QUESTION

2. Give three examples of management controls in relation to noise.

(Suggested Answer is at the end.)

Summary

This element has dealt with the control of environmental noise.

In particular, this element has:

- Described noise as unwanted sound. Low-frequency sounds can travel long distances and have been associated with ill-health effects, such as nausea, headaches and anxiety. Other nuisances include tannoy systems and uncontrolled sirens.
- Discussed sources of industrial environmental noise, such as:
  - Noise from commercial activities, including machinery, extraction systems, compressor systems and public address systems.
  - Transport noise, e.g. engine noise, use of horns and reversing signals.
  - Agricultural noise, e.g. complaints relating to tractors, machinery and bird-scarers.
  - Construction noise, e.g. erection, construction, alteration, repair or maintenance of buildings, structures and roads.
  - Quarrying and mining.
- Outlined other sources of noise, including:
  - Noise from pubs and clubs.
  - Noisy neighbours.
  - Intruder and vehicle alarms.
- Described methods for the control of environmental noise, including:
  - Elimination, substitution (replace diesel/petrol engine with electric motor) and maintenance (replacement of worn parts/regular oiling).
  - Isolation, by enclosing the noise source.
  - Absorption and insulation, by using sound-absorbing materials on walls and other large surfaces.
  - Environmental noise barriers, placed in the right position.
  - Damping, by using anti-vibration dampers or rubber mountings.
  - Silencing, by the use of silencers which absorb the sound pressure generated by the process at its source.
  - Management controls, e.g. control of working hours, use of radios, public address systems, vehicle routes, loading doors and shutters.

Exam Skills

Question

Scenario

A company that manufactures wooden furniture is planning to renovate its production facility. This will involve the installation of new automated machinery to cut and shape wooden materials that is currently undertaken at another site. It is planned that the new facility will operate on a two-shift schedule with some night-time working involved. The factory's southern boundary is adjacent to residential housing.

Task

A fan and collector hopper is planned to be used to collect wood dust and is to be installed against an outside wall of the factory.

(a) What are the potential sources of noise from the fan and hopper? **(4 marks)**

(b) What are the issues that should be addressed to ensure that the fan and hopper do not cause a noise nuisance to the local community? **(6 marks)**

Approaching the Question

Think about the steps you would take to answer the question:

- Read the scenario carefully. With this question you need to (a) consider the exact source of noise from the fan and hopper and (b) the control and other measures to eliminate complaints.

- Now look at the task – prepare notes on:
  - The constituent parts of a fan or hopper that may cause noise.
  - Methods to control the noise from the relevant parts of the fan and hopper.

- Consider the marks available. In this case, there are 4 marks available for part (a) so you should provide at least 4 noise sources. There are 6 marks available for part (b) so you should provide 6 methods to reduce the risk of nuisance noise.

- Read the scenario and task again to make sure you understand them and have a clear understanding of fans/hoppers, how they can create noise and how the noise could be controlled.

- Jot down an outline plan - this might include:

  (a) The motor and fan; vibration of the equipment, including grills and ducting; the discharge point.

  (b) Equipment choice, maintenance, location, time of day, attenuation techniques, monitoring.

Now have a go at the question yourself.

Exam Skills

Example of How the Question Could be Answered

(a) The potential sources of noise from the fan and hopper include the motor and fan - they may not have been properly maintained and require fixing or replacement. There is also a significant potential for parts of the system to vibrate - ducting for example has a significant potential to cause vibration. Noise may also occur from the discharge point.

(b) The correct equipment for the task should be chosen. This would include an evaluation of the noise levels of the whole system, taking into account proximity to sensitive receptors.

The fan and collector hopper would also need to be subject to regular preventive maintenance. Lack of lubrication or worn bearings or other parts could be a significant noise source.

The location of the equipment will also be important. Locating the equipment away from sensitive receptors, such as residents of surrounding housing, will result in less chance of nuisance occurring.

The time of day of the operation of the unit will also be important. The fan is more likely to cause a nuisance if operated at night, due to low levels of background noise and the increased likelihood of people being in their homes.

The noise could be attenuated by using techniques such as sound insulation, vibration mountings and screening by enclosures.

A regular monitoring regime could also be implemented. This would also assist in determining whether the unit is the actual source of the noise, should a complaint arise.

Element 7

Control of Contamination of Water Sources

Learning Outcomes

- Understand the importance of reducing environmental harm; identify sources of water pollution; and suggest suitable control measures.

Learning Objectives

Once you've read this element, you'll be able to:

1. Demonstrate awareness of the environmental impacts of water pollution.

2. Identify sources of environmental harm and suggest suitable control measures for emissions.

Contents

Importance of the Quality of Water for Life — 7-3
The Meaning of Safe Drinking Water, Groundwater, Surface Water — 7-3
The Water Cycle — 7-5
Water for Agriculture and Industry — 7-7
The Potential Effects of Water Pollution to the Environment — 7-7
Over-Abstraction — 7-8
Desalination — 7-8
Water Conservation — 7-8
The Potential Effects of Pollution on Water Quality — 7-9
Main Issues and Impacts of Ocean Pollution — 7-9

Main Sources of Water Pollution — 7-11
Controlling Sources of Water Pollution — 7-11

Main Control Measures Available to Reduce Contamination of Water Sources — 7-17
Control Hierarchy — 7-17
Monitoring Water Quality — 7-18
Control Methods — 7-19
Controls for Storage and Spillage — 7-20
Controls for Wastewater — 7-26
Difficulties in Maintaining Equipment in Some Locations or Environments — 7-28

Summary — 7-29

Exam Skills — 7-30

Importance of the Quality of Water for Life

IN THIS SECTION...

- Drinking water is sourced from groundwater, reservoirs and rivers. It is treated to provide an adequate and continuous supply of water free from pathogens and other undesirable characteristics.
- Water is continuously transported around the water cycle, in either liquid, vapour or ice.
- It is important we protect groundwater and rivers as they are an essential resource.
- Water conservation is important as less than 1% of the water on the planet is available for use.
- Pollution and over-abstraction of water and groundwater can affect human health and impact ecosystems.

The Meaning of Safe Drinking Water, Groundwater, Surface Water

Water is essential for life:

- We need clean, fresh water to drink and for washing.
- Our food crops need water for irrigation.
- Rivers, lakes and the seas provide habitats for wildlife and provide us with food and places of recreation.

Safe drinking water is defined by the World Health Organization (WHO) as being: "*water that does not represent any significant risk over a lifetime of consumption, including different sensitivities that may occur between life stages*".

Waterborne diseases, transmitted by poor-quality drinking water and lack of sanitation, are among the most important health challenges facing mankind. The World Health Organization estimates that:

- Some 1.6 million people - mostly children under the age of 5 - die every year from diarrhoeal diseases (e.g. cholera).
- More than 100 million people suffer from intestinal helminth (parasitic worm) infections.

Natural water

DEFINITION

PATHOGENS

Disease-causing organisms, such as bacteria and parasites, that cause diseases such as cholera, typhoid, dysentery, bilharzia and hookworm.

As well as carrying a large number of pathogens, drinking water may have a range of other undesirable characteristics:

- Unpleasant colour, e.g. due to dissolved organic matter.
- Turbidity, e.g. caused by suspended mineral or organic matter.
- Unpleasant taste and smell, e.g. due to sewage contamination.
- High mineral content, e.g. minerals absorbed from contact with soil, such as calcium or magnesium sulphates, that cause 'hardness'.

Safe drinking water is water with microbial, chemical and physical characteristics that meet WHO guidelines, or national standards on drinking-water quality.

7.1 Importance of the Quality of Water for Life

Water treatment is required to produce an adequate and continuous supply of safe drinking water. In general terms, this means water that is:

- Free from pathogens.
- Free from harmful mineral content.
- Clear, i.e. is not turbid or coloured.
- Palatable, i.e. has no unpleasant taste.

Surface water incorporates all water found on the surface of the Earth. It includes freshwater located in rivers, streams and lakes and saltwater present in the oceans. Surface water may persist for a limited period or may be permanently present.

Groundwater consists of water that is found below the surface in soil and geological formations such as sandstone and limestones. Groundwater is an important source of drinking water and water for rivers and other watercourses. As groundwater is hidden away, the effects of pollution cannot be immediately seen. Because groundwater is vulnerable to pollution and can be easily damaged, there are often specific policies and laws in place to protect it.

Groundwater pollution usually occurs gradually. The sources of groundwater pollution often include leaks from underground storage tanks and wash-off of contaminated rainwater. It can also be difficult to detect as it may move very slowly through porous rocks but in fissured aquifers it can move much faster.

> **MORE...**
>
> The WHO *Guidelines for drinking-water quality* define a framework for drinking water safety, including limits for potential pollutants:
>
> www.who.int/water_sanitation_health/publications/2011/dwq_guidelines/en

Importance of the Quality of Water for Life | 7.1

The Water Cycle

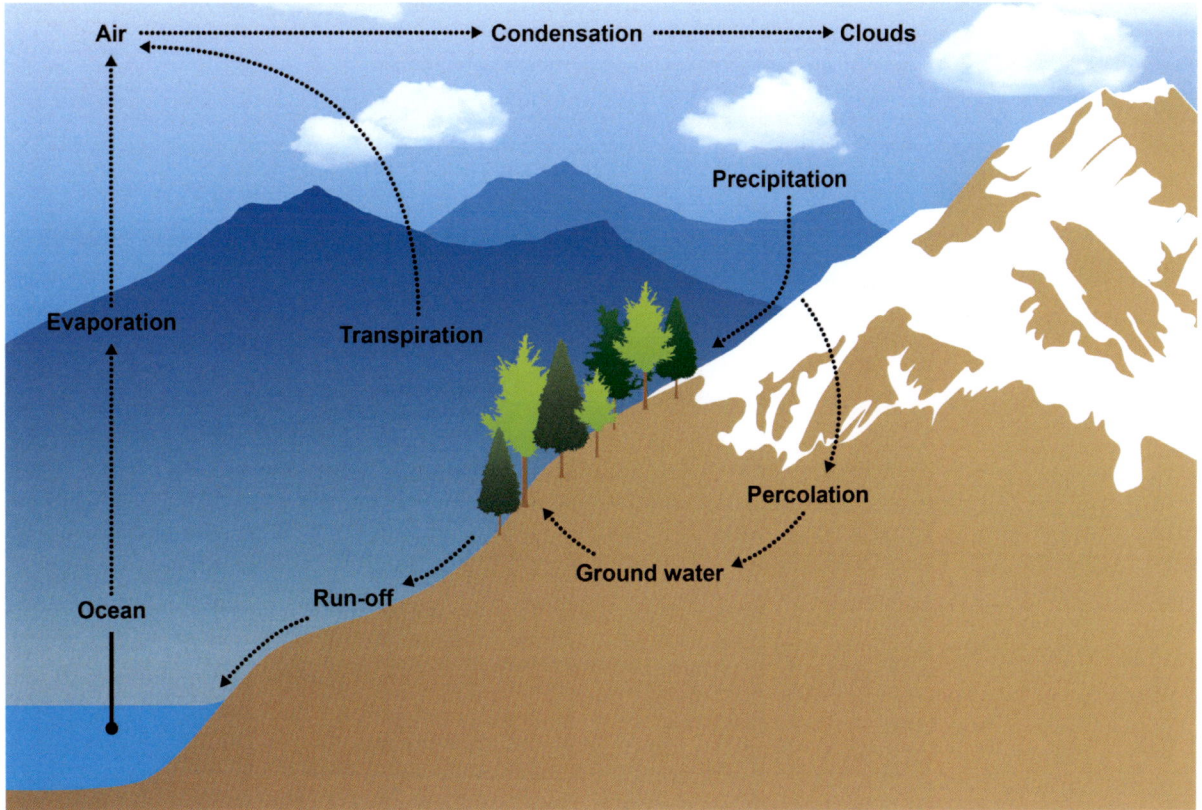

The water cycle

The **water cycle** (see diagram) is unique in that water is present throughout only as the molecule H_2O, albeit existing in three physical states - vapour, liquid and ice. It is not chemically transformed.

Liquid water **takes in latent heat energy** to become water vapour; and water vapour condenses to liquid water, **releasing latent heat energy**. The amounts of energy involved are very large and the dynamics of weather are in great measure driven by them.

> **DEFINITION**
>
> **TRANSPIRATION**
>
> The process of water movement through a plant and its evaporation from the plant surface to atmosphere.

Water moves around the Earth through a system known as the **hydrological cycle**. For water to complete the full cycle it can take thousands of years.

- The initial input of water in the system is in the form of precipitation, which either seeps into the land surface (percolation), or runs over the surface.
- The amount of water that will run off will depend on the permeability of the ground and the catchment area. If conditions are dry, more water will seep in, but after heavy rain the ground can become saturated, resulting in more run-off. Run-off may be greatly increased in urban areas, which can lead to flooding if the drainage systems do not have sufficient capacity.
- Plant roots can take up water that has seeped into the soil. If the water contains pollutants, they can be drawn up into the plant and released (transpiration) and possibly transferred to another natural cycle, i.e. if eaten by animals or humans.
- The water can continue to seep through the soil horizons to reach aquifers (water-bearing rocks) and form part of the groundwater supply.

7.1 Importance of the Quality of Water for Life

As these processes are happening, the power of the Sun is driving this cycle by causing evaporation. This is the change of liquid water to a vapour. Sunlight aids this process, as it raises the temperature of liquid water in oceans and lakes. As the liquid heats, molecules are released and change into a gas. Warm air rises up into the atmosphere and becomes the vapour involved in condensation. Because of this cycle, there can be an accumulation of pollutants through water catchments, making prevention of pollution particularly important.

Although there appears to be a vast abundance of water available on the planet, most of this is seawater. Only a small fraction of it is freshwater that is readily available to us for drinking, industry, agriculture, etc.

Distribution of water across the planet

Location	% of Total
Oceans	97.24
Glaciers and icecaps	2.14
Groundwater aquifers	0.61
Lakes (freshwater)	0.009
Inland seas	0.008
Moisture held in soil	0.005
Atmospheric moisture	0.001
Rivers	0.001
Total	100

Drinking water is typically sourced from:

- Rivers and lakes, known collectively as 'surface waters'.
- Groundwater, e.g. via springs, wells and boreholes.
- Artificial reservoirs, e.g. created by building dams across river valleys.
- Desalination (see later) - salt may be removed from saltwater and the water made safe to drink.

Drinking water may also be obtained from the desalination of seawater, but this is highly energy-intensive and expensive.

Being such a valuable and essential resource, water is:

- Continuously re-used and recycled and great attention is paid to protecting rivers and groundwater.
- Often vigorously protected by criminal law, with significant penalties available to the courts for anyone who pollutes a source of drinking water.

Water sources are often controlled by stringent legislation around the world. In many countries, it can be an offence to discharge into the following types of watercourse without a permit or other legal authorisation:

- Relevant territorial waters.
- Coastal waters.
- Inland freshwaters.
- Groundwater.

For example, permits are required in England and Wales under the **Environmental Permitting (England and Wales) Regulations 2016** where anyone wishes to discharge into the above waters.

Importance of the Quality of Water for Life | 7.1

Water for Agriculture and Industry

Agriculture (70%) and industry (22%) account for the greatest proportion of all of the freshwater abstracted from surface and groundwaters worldwide. The proportion of water used by households for drinking and washing is comparatively small (8%).

Large quantities of water are used for irrigating crops - although the actual amount varies considerably between different regions and climates. A large quantity of water is also used for farm animals (both directly for drinking and also indirectly through the food they eat). For example, about 15m^3 of water is typically required to produce one kilo of beef.

Certain industries also consume significant amounts of water. Water may be incorporated directly into food and drink products, and also some chemical products (e.g. water-based paints). Water may also be used in manufacturing as process water, especially for cooling (e.g. power stations, steel production) and for cleaning.

The Potential Effects of Water Pollution to the Environment

There are a number of critical ways in which pollutants can have an adverse impact on the environment.

Physical Impacts

Plants and animals that live on the bottom of rivers and lakes may become physically smothered by effluents that contain a high concentration of solids. Untreated sewage, run-off from mining, quarrying and construction activities may all contain high levels of suspended solids, such as silt, sand and organic particles.

High levels of suspended solids are also likely to make the water very turbid, reducing the light available for aquatic plants to grow. This will have a knock-on effect on animals that feed on the plants.

Oxygen Stress

An oil spill can be devastating for local wildlife

> **DEFINITION**
>
> **MICRO-ORGANISM**
>
> An organism that is microscopic, including bacteria, fungi, microscopic plants and animals such as plankton.

Plants and animals that live in aquatic environments depend on an adequate supply of oxygen for respiration - just like plants and animals on land. But land plants and animals get their oxygen direct from the atmosphere, whereas aquatic plants and animals must use oxygen that is dissolved in the water in which they live. Although oxygen makes up around 21% of the atmosphere (by volume) it is much less available in water. The amount of oxygen that water can hold in solution depends on a number of factors, especially temperature, but typically a litre of water will only contain 5-6 milligrams of oxygen - this is around 20 times less than a litre of air. Aquatic plants and animals are therefore extremely sensitive to any pollutant that reduces the oxygen content of the water.

Creating oxygen stress is one of the most important mechanisms by which polluting effluents can affect aquatic wildlife. Effluents that contain high levels of organic material - e.g. human sewage - strip oxygen from receiving waters. This is because oxygen is used in the breakdown of organic material which increases as they grow and multiply.

Eutrophication - a process of nutrient enrichment - is a related process that also results in oxygen stress in aquatic environments. This is caused by excessive nutrients (e.g. run-off of fertilisers from agricultural land) greatly enhancing the growth of aquatic plants, especially microalgae that live suspended in the water column. As the plants grow and multiply, they also consume the oxygen in the water, greatly reducing the oxygen that is available for animals such as fish.

7.1 Importance of the Quality of Water for Life

The 'blooms' of microalgae created by eutrophication may also reduce the light available for plants that live on the bottom of the lake or river, restricting their growth.

Toxics

A wide range of contaminants that may be present in effluents are potentially poisonous to aquatic wildlife. These include toxic metals such as copper, mercury and lead, spills of fuel and oil, and chemicals, such as solvents. Biologically-active materials, such as insecticides and herbicides, pose a particular threat if they contaminate aquatic habitats.

Over-Abstraction

Excess demand for water, leading to over-abstraction from water sources, can also have detrimental impacts on aquatic wildlife. Water removed from ground and surface waters rarely returns to the source from which it has been taken. Around 80% of water applied to crops evaporates, for example. This can lead to many rivers flowing at low levels during times of peak demand or, in the worst case, they can dry up completely. Impacts of over-abstraction on rivers and aquifers include:

- Reductions in river-water flow reducing the size of the populations of aquatic species that the river can support.
- Wetland habitats that are supported by river flows drying up and disappearing.
- Aquifers drying up, removing important sources of water for human consumption and agriculture.

In coastal areas, removing water from aquifers at an excessive rate can lead to saltwater intrusion, making the water unfit for use.

Desalination

Desalination is the removal of minerals, largely salt, from saline water. It produces water that is suitable for human consumption and agriculture, and salt as a by-product. The need for additional freshwater is great in arid areas that have limited access to surface or groundwater.

Saline water can be desalinated in various ways such as:

- Multiple stage flash distillation.
- Vapour compression distillation.
- Reverse osmosis.

The process to desalinate water requires significant amounts of energy and associated impacts are great. Distillation, for example, requires large amounts of water to be boiled and reverse osmosis requires significant energy to overcome natural osmosis. Desalination plants are generally expensive to build and maintain. It has been found that desalinated water can be five times as costly as freshwater. Other environmental impacts, in addition to those associated with energy, can be great. Chlorine and other chemicals are often added to water during processing and left behind with waste brine which, if dumped into the ocean, can harm marine life.

Water Conservation

With less than 1% of the water on the planet actually available for use, water should be treated as a valuable resource. Even in countries where it is comparatively readily available, we should make an effort to conserve water where possible. This conservation also has a direct and positive effect on energy savings, as energy is used throughout the process that brings water to our taps.

Some of the ways to conserve water include:

- Toilets - if installing a new toilet, ensure it has a dual flush system which allows less water to be used if a full flush is not required. Indeed, consider not always flushing the toilet; even a short flush system uses several litres of clean water and may not always be necessary. If you have the older, single-flush system, then consider a water-saving device, such as a 'Hippo' - a plastic container open at the top that retains a portion of the water that would have been used in the flush.

Importance of the Quality of Water for Life   7.1

- Fit a water meter - knowing you are being charged for what you use is a great incentive to reduce water consumption. It can also save you money on both your water bill and sewerage bill as this is calculated from the amount of water you use.
- Stop dripping taps - according to Waterwise (a not-for-profit UK water organisation funded by the water industry), a dripping tap wastes at least 5,500 litres of water a year.
- Water garden plants in the evening - this ensures that more of the water remains available to the plants and so in the long run less has to be used.
- Fit diffusers on taps - they won't make much difference when filling a bowl or basin, but if you wash anything under a running tap they will reduce the amount of water needed.
- Grey water recycling - using bath and washing water to flush the toilet can save large quantities of fresh, clean, drinking water from simply being flushed away.
- Fit low flow showerheads and take more showers than baths.

Balancing the Different Needs

Water needs must be balanced between what is required by nature and what is needed for human activities (e.g. agriculture and industry).

The Potential Effects of Pollution on Water Quality

Drinking contaminated water may affect human health in a variety of ways, depending on the concentration and nature of the contaminant.

Environmental enforcement bodies often assess and classify the quality of rivers and other water bodies. The assessment method can look at both the ecological (considering fish and invertebrate species present) and the chemical status (e.g. concentrations of pollutants such as heavy metals, pesticides and nutrients) of the water. The classification system identifies where the water quality is good or where it needs to be improved.

One common ecological method of classifying river quality uses invertebrate species as a basis for measurement. Known as the Biological Monitoring Working Party (BMWP) score, it attaches a score between 1 and 10 to species of aquatic invertebrates depending on their tolerance to pollution (the less tolerant a species is, the higher the score). Sensitive species, such as stonefly nymphs, attract a score of 10, while more tolerant species, such as worms, have a much lower score. By using a simple hand net, a sample can be obtained and examined and scores given for the number of species found in the sample. (Note that scores are for number of species, not number of individuals found, so five stonefly nymphs still attract a score of 10, as would one stonefly nymph.)

As we saw in Element 1, excessive levels of nitrates and phosphates in relatively still waters, such as lakes, can lead to **eutrophication**.

Main Issues and Impacts of Ocean Pollution

Most of the pollutants that enter the ocean result from human activities either inland or on the coast. They may occur from point sources, such as discharges of sewage, or non-point sources such as agricultural run-off. A significant problem with ocean pollution is discharge of nitrates and phosphates. As we considered earlier, these enter coastal waters, causing algal blooms; the algae are then decomposed and strip oxygen from the water, depleting the supply for marine life, causing death or mobile species to vacate the area.

Marine debris is also a significant oceanic pollution problem. The world's oceans are polluted by numerous types of solid items ranging from microplastic materials, to larger items such as fishing lines. Some plastic materials can take over 400 years to degrade. Many marine species have been negatively impacted by such pollution, as it can cause mortality or serious harm by entanglement or ingestion. In fact there is so much waste material in the world's oceans that there are five 'garbage patches' where large quantities of waste material have accumulated in gyres (large systems of rotating ocean currents).

Oceans also absorb large quantities of carbon dioxide from the atmosphere. An excess of these gases is causing oceans to become more acidic and as such threatens the habitats of marine species, with coral and many types of plankton being affected.

7.1 Importance of the Quality of Water for Life

> **MORE...**
>
> *How to manage change and improve water efficiency* (WRAP guide)
>
> Provides information on how to assist companies in establishing changes that improve water efficiency. You can find it at:
>
> www.wrap.org.uk/sites/files/wrap/Behavioural%20change%20handbook.pdf

> **STUDY QUESTION**
>
> 1. Virtually all water bodies, such as rivers, lakes and groundwater, are protected by criminal law. Explain why it is important that all types of water body are protected.
>
> (Suggested Answer is at the end.)

Main Sources of Water Pollution

IN THIS SECTION...

- Water pollution can be caused by 'point sources' or 'diffuse sources'.
- Many sources of water pollution need to be controlled. These are: domestic wastewaters (sewage), surface-water drainage, industrial discharges, process and cooling water, mining and quarrying, litter, agriculture, contamination from natural materials and unplanned discharges.

Controlling Sources of Water Pollution

We have highlighted the need to conserve our precious water resources by using water wisely and not polluting the water that is available to us.

The main sources of water pollution that we need to control are discussed below.

Domestic Wastewaters - Sewage

In the 1800s, populations began to rapidly increase - particularly within major cities. Untreated human waste was discharged into local rivers, resulting in an awful smell. This was a common issue across the world.

Although the network of underground sewers in most cities has dramatically improved the situation, domestic waste is still the greatest potential source of water pollution around the globe. This waste is often referred to as **sewage**.

Sewage comes from many sources:

- Domestic premises - toilets, sinks, baths and showers, dishwashers, washing machines, etc.
- Commercial premises, e.g. offices, industry, hotels and restaurants.
- Rainwater - which runs from roads, pavements, roofs, etc. into drains which may also link to the sewers.

Untreated human waste discharged into a local river

In the UK, we dispose of 11 billion litres of sewage every day. Historically, a very small proportion of this sewage would have been treated (if any). However, there are now international standards in place which control how sewage should be treated before being discharged into rivers or oceans.

It is still the case though, that domestic wastewater from many cities around the world - especially in developing countries - is still discharged untreated. This is a major source of water pollution.

7.2 Main Sources of Water Pollution

The main pollutants in sewage are:

Pollutant	Potential Impacts
Solid debris: • Plastic waste. • Wood. • Textiles.	• Litter on the surface of water and around river banks and shorelines.
Organic material: • Human waste. • Food residues.	• Using up oxygen as it decays - this is known as Biological Oxygen Demand (BOD). Can strip enough oxygen from the receiving water so that fish and other wildlife are unable to survive.
Suspended solids -- silt, sand and organic particles suspended in the water.	• Smothering plants and animals that live on the bottom of the receiving water. • Clouding the water, reducing the light available for plants to grow. • Toxic contaminants, disease-causing viruses and bacteria may adhere to these particles, which may in turn be eaten by aquatic animals.
Nutrients -- nitrogen and phosphorus compounds.	• Promoting excessive growth of microscopic plants, which strip oxygen from the water.
Toxics - metals (such as copper, nickel and lead) and oil.	• Poisoning of wildlife and humans (if quantities are high enough).

Surface-Water Drainage

As we have seen, rainwater that runs off surfaces such as pavements and roads may drain into the sewer system and be treated with domestic sewage. However, many sources of surface drainage, e.g. from industrial installations, construction sites, quarrying and mining activities, may bypass the sewer network. This is a very important potential source of pollution of both surface and groundwaters.

Rainwater that falls on surfaces will collect any contaminants that may be present. The range of contaminants will depend on the activities being undertaken on the site, but is likely to include dirt and other solids and residues of oil, raw materials and process chemicals that may be handled and spilt on site. There may therefore be a wide range of potentially hazardous materials entering surface water drains. These drains may discharge directly into surface waters, e.g. the local river (with or without treatment), or drain into a soakaway buried in the ground where contaminants may come into contact with groundwater.

Grit and silt from construction activities can run off into rivers and lakes

On many industrial sites, especially older sites, the surface drainage system may be poorly understood and the final points of discharge may not be fully identified. This makes pollution incidents more likely, e.g. in the event of a spill entering the surface water drainage system.

Spillage onto uncovered ground can cause land contamination. Additionally, the land can act as a pathway for the pollutants to be transferred to controlled water such as a river or stream. Many small spills occurring over a long period of time can lead to build-up of large quantities of contaminants in the land which can subsequently be passed to watercourses.

Industrial Discharges

Liquid effluents are by-products of many industrial processes and industries. Many companies discharge their liquid waste into the existing sewage system, where it will be processed through a sewage treatment works before being discharged. However, some larger companies operate their own sewage treatment works and discharge directly into local waters in accordance with relevant legal standards. Unfortunately, in some parts of the world, the treatment stage is not required, or not enforced.

The operating processes of the following industries can produce huge volumes of liquid effluent:

Industry	Water-Pollution Issues
Textiles	Washing, dyeing and rinsing fibres generates large volumes of wastewater. This wastewater is often contaminated with a range of chemicals, including traces of pesticides used on raw materials.
Pulp and paper manufacture	Pulping timber to make paper requires large quantities of water and results in an effluent with high levels of BOD and suspended solids. Highly-toxic residues of chlorine from paper bleaching may also be present.
Manufacture and bottling of soft drinks and alcoholic beverages	Cleaning equipment and bottles and disposing of waste product generates large volumes of wastewater with a high BOD.
Milk and associated dairy processing	Cleaning equipment and bottles and disposing of waste product generates large volumes of wastewater with a high BOD.
PVC production (chloralkali processes)	Traditional chloralkali processes use mercury electrodes, meaning effluents may contain highly-toxic mercury compounds.
Titanium-dioxide production	This white pigment is extracted from natural minerals in a process that creates large quantities of acidic liquid sludge.

Process and Cooling Water

In some industries, water is used for cooling plant and equipment. Although coolant water is usually largely free from contaminants, it will have become hot. Discharging coolant water before it's cooled can cause the temperature of a river to rise by more than 10°C. Warmer water is able to hold less oxygen, meaning local wildlife can suffer as a result. Permits are often required to ensure the discharge of coolant water is properly managed and controlled.

Sectors which require a lot of water for cooling include: electricity generation; oil refining; steel-making; cement manufacture; and paper manufacture.

Mining and Quarrying

Mining, quarrying and associated ore-processing activities have the potential to create serious water pollution. Run-off is a particularly important source of contamination that may wash into nearby watercourses, or seep into groundwater.

Run-off from mining and quarrying is likely to contain high levels of solids. As we have seen, discharges that contain excessive solids may smother aquatic life and reduce levels of light available for photosynthesis in the receiving water. Run-off may contain solids from:

- Extractive operations that use water, e.g. hydraulic processes to extract china clay.
- Mineral washing and processing activities.

7.2 Main Sources of Water Pollution

- Water drained from excavations to allow work to continue ('dewatering').
- Spoil heaps and stockpiles.
- Washing vehicles, work areas and equipment.

Oil and fuel spills are also an important source of contaminants in run-off from mining and quarrying sites.

Certain types of mineral extraction, e.g. the excavation of coal and metal ores, may expose rocks that contain sulphur. If water from the workings comes into contact with this sulphur, it can result in the formation of sulphuric acid. The resulting effluent is known as Acid Mine Drainage (AMD).

In addition to its acidic nature, AMD may contain potentially toxic metals in solution, such as lead, zinc, iron, mercury and cadmium. AMD is often a particular concern when mine workings are closed down, because disused workings will tend to flood when active dewatering (i.e. pumping out) of the workings ceases, allowing exposed sulphur to mix with water.

An acidic run-off may also arise from spoil heaps and stockpiles of crushed and waste rock, known as Acid Rock Drainage (ARD), which can cause similar problems to AMD.

Solid Materials

Significant amounts of litter, especially plastics such as old bottles and wrapping, end up in our rivers, lakes and on beaches. As well as being unsightly, old containers may contain residues of oil, fuel and other potentially hazardous materials.

Surface drainage systems can be an important route by which litter reaches the environment. Construction sites may also generate significant amounts of litter, and other solids, such as grit and cement dust, may be exported in run-off.

Agriculture

Agricultural operations are often associated with potential water pollution:

- Hazardous liquids, such as fuel and pesticides, are often stored on site.
- Animal waste may be stored in slurry pits.
- Large volumes of milk may be stored.
- Silage and other effluent from feed stores may be a source of water pollution.

Generally, agricultural pollution results from an unplanned release whereby these contaminants make their way into rivers - often through spillage or problems with storage.

Pesticide application may lead to run-off into a watercourse

The run-off of fertilisers and pesticides from fields is the most significant cause of water pollution associated with agriculture. Fertilisers are generally rich in nitrates and phosphates and, if used in inappropriate quantities, locations or times of year, these fertilisers can make their way into drainage systems and rivers. Because this type of pollution does not reach the river through a distinct source, it's known as a **diffuse source**.

Raising the levels of nitrates in water can result in excessive growth in microscopic plants which remove oxygen from the water - meaning other wildlife is endangered. It would also be dangerous where the river water is used for drinking supplies because soluble nitrates are not always removed by water purification processes.

However, milk also presents a serious threat to the environment; if it enters a river its high BOD can also be devastating.

> **MORE...**
> NetRegs has produced useful guidance on preventing water pollution. It is available at:
>
> www.netregs.org.uk/environmental-topics/water/preventing-water-pollution

Contamination from Natural Minerals

Not all pollution comes from man-made sources; naturally-occurring minerals may also result in significant impacts. Two examples of natural pollutants are highlighted in the following cases:

- In Bangladesh, the main source of drinking water is groundwater. Arsenic has been identified in groundwater in many locations in the country and has caused widespread concern. Arsenic is a metal known to be a carcinogen (as well as having other health effects) and is soluble in water. Arsenic is present because it occurs naturally in the rocks through which groundwater flows.

  It is estimated that between 25 and 36 million people are exposed to arsenic levels that exceed the Bangladesh standard of 50 ppb (parts per billion). Tens of thousands of people have been identified as suffering from skin discolouration and other more serious complications of arsenic toxicity.

- Radon is a colourless, odourless, tasteless, naturally-occurring radioactive gas that is associated with igneous rocks, such as granite, that contain uranium. Radon gas may migrate to the ground surface and then penetrate and accumulate in buildings. Exposure to high levels of radon gas in the atmosphere increases the long-term risk of lung cancer.

  Large parts of Cornwall (a county in the south-west of the UK) are classed as radon-affected areas because 1% of domestic properties have a radon level above the safe limit of 200 becquerels of radioactivity per cubic metre of air (Bq/m^3).

  The most important route for radon to contaminate the atmosphere inside homes is through the direct migration of the gas from the underlying ground. However, in radon-affected areas, drinking water that is obtained from groundwater sources may also contain very small amounts of dissolved radon gas. This dissolved gas may be released from the water supply through activities such as showering, increasing levels of the gas in the indoor atmosphere.

Unplanned Discharges

Many of the sources of water pollution that we have discussed so far result from planned activities, such as point-source discharges from industrial processes. Discharges from planned activities such as these are usually carefully controlled to comply with legal requirements, especially the conditions of any discharge permit.

But water pollution from unplanned activities can also be very significant. Accidental spills and leaks of fuel and liquid chemicals are the most common cause of unplanned discharges that can cause water pollution. Spills often happen when tanks are being refilled or decanted. Old or disused tanks may also leak if they become damaged or corroded. Contaminating materials that occur on unmade ground (i.e. bare or unprotected ground) may be particularly difficult to control, because contaminants can migrate considerable distances through porous ground, such as sands and gravels, and eventually pollute groundwater a long way from the point of the spill.

While many spills are minor, some unplanned incidents can have massive environmental and economic repercussions. The Deepwater Horizon oil spill that took place in 2010 in the waters of the Gulf of Mexico was caused by an explosion on an oil-exploration rig. In all, 780,000m^3 of crude oil were discharged into the ocean, affecting marine and coastal ecosystems for hundreds of miles and resulting in more than US$50 billion costs for BP, the operator of the project.

Unplanned discharges may often result in diffuse pollution. River pollution from the run-off of fertilisers and pesticides from agricultural activities (see above) is an important example of an unplanned discharge.

7.2 Main Sources of Water Pollution

STUDY QUESTIONS

2. State the two main categories of water pollution sources.
3. List any three of the main sources of water pollution.

(Suggested Answers are at the end.)

Main Control Measures Available to Reduce Contamination of Water Sources

IN THIS SECTION...

- Contamination of water can be reduced by considering the control hierarchy.
- Permits to discharge and groundwater pollution are covered by certain EU Directives.
- Reactive and active (proactive) methods can be used to monitor contamination of water sources.
- Physical measures to prevent or reduce pollution to water include:
    - Bunding of stores.
    - Use of oil interceptors.
    - Spill response procedures.
    - Coagulation to remove solids.
    - Correction of pH and temperature.
    - Screening, sedimentation, filtration and centrifugal separation to remove solids.

Control Hierarchy

You should be aware of the hierarchical duty to 'eliminate, minimise and render harmless' emissions to the environment, which we covered earlier.

> **TOPIC FOCUS**
>
> **Control Hierarchy for Water Pollution**
>
> - **Eliminate:**
>   - Replace chemicals that are harmful to the aquatic environment with non-hazardous alternatives.
>   - Change of process to produce a solid rather than a liquid waste.
> - **Minimise:**
>   - Reduce the amount of water used in a process or activity.
>   - Store smaller quantities of hazardous substances at any one time.
>   - Reduce the amount of fertilisers used on agricultural land.
> - **Render harmless:**
>   - The techniques described in the subsection on **Control Methods** in this element either minimise pollutants or render them harmless before they are discharged to water.

7.3 Main Control Measures Available to Reduce Contamination of Water Sources

Monitoring Water Quality

Permit conditions may include specific monitoring (for water quality) and maintenance requirements.

> **TOPIC FOCUS**
>
> **Active and Reactive Monitoring**
>
> Active monitoring is undertaken before there has been a failure. Examples would include:
>
> - Sampling the quality, flow rate, pH and other parameters of the water discharge.
> - Mass balance calculations for underground storage tanks.
> - Site inspections to identify potential risks.
> - Calibration of monitoring equipment to ensure accurate results.
>
> Reactive monitoring is undertaken following a failure. Examples would include:
>
> - Collecting data on near misses.
> - Monitoring of complaints from neighbours or workers.
> - Information on enforcement action.
> - Records of past incidents or spillages.
>
> (Note: similar active and reactive monitoring is appropriate for emissions to air.)

Monitoring should include a mixture of active (proactive) and reactive measures. Active monitoring is undertaken before there has been a failure. Examples would include:

- Sampling the quality, flow rate, pH and other parameters of the water discharge.
- Mass balance calculations for underground storage tanks.
- Site inspections to identify potential risks.
- Calibration of monitoring equipment to ensure accurate results.

Active monitoring may be undertaken of both the quality of a process effluent stream prior to discharge into a receiving water, and the quality of the receiving water itself. In either case, the variables that are likely to be monitored are those that are typically included in permits to discharge:

- Flow rate.
- Temperature.
- pH.
- Concentration of suspended solids.
- Chemical Oxygen Demand (COD)/Biological Oxygen Demand (BOD)/Total Oxygen Demand (TOD).
- Concentration of dissolved oil.
- Concentration of dissolved metals.

Other tests for specific contaminants (such as a specific pesticide) may be undertaken if these are expected to be present or are of particular interest.

Reactive monitoring is undertaken following a failure. Examples would include:

- Collecting data on near misses.
- Monitoring of complaints from neighbours or workers.
- Information on enforcement action.
- Records of past incidents or spillages.

(Note: similar active and reactive monitoring is appropriate for emissions to air.)

7.3 Main Control Measures Available to Reduce Contamination of Water Sources

Control Methods

We will now look at some of the legal and physical controls available to reduce pollution of water resources.

Permits/Licences to Discharge

Discharges of wastewaters to sewers, rivers, lakes and other watercourses must be within limits set by an enforcement body. Often, a permit to discharge is used as the key legal control in many jurisdictions. Parameters that are set in permits can vary but may include:

- Maximum permitted flow rate (daily and hourly).
- Temperature.
- Maximum Chemical Oxygen Demand (COD) or maximum Biological Oxygen Demand (BOD). These and related terms are described in more detail below.
- pH range (typically 5-9).
- Maximum concentration of suspended solids.
- Limits of amounts of dissolved oil, metals (e.g. copper, zinc), organic chemicals (e.g. phenols).
- Limits on pesticides and heavy metals (e.g. cadmium, mercury and lead).

You should also be aware that in the EU, limits for discharges to water can be set under an integrated permit as required by the **Industrial Emissions Directive (2010/75/EU)**.

Groundwater Protection

As a result of its importance, groundwater is often protected by law. In the EU, for example, **Directive 2006/118/EC** (the **Groundwater Directive**) provides control on hazardous substances and non-hazardous pollutants that could contaminate groundwater.

Table identifying examples of groundwater pollutants (the substances shown are those identified in the Environmental Permitting (England and Wales) Regulations 2016)

Groundwater Hazardous Substances	Groundwater Non-Hazardous Pollutants
- Heavy metals - Pesticides - Radioactive substances - Hydrocarbons - Discharges from septic tanks	- Ammonia - Metals - Biocides

> **MORE...**
>
> Further information on groundwater protection can be viewed at:
>
> www.gov.uk/government/publications/protect-groundwater-and-prevent-groundwater-pollution

Water Quality Indicators

COD, BOD and TOD are methods of measuring the potential oxygen depletion that can be caused following discharge of pollutants into water. This occurs from the breakdown of organic materials by micro-organisms which subsequently take oxygen out of the water as part of the process of decomposition. Such oxygen depletion can severely affect aquatic life, killing fish for example (fish do not have enough oxygen to breathe). Substances that cause such pollution include milk, beer, sewage, blood, etc. As such, all have to be discharged within consent conditions (obviously, this may also be a legal requirement) and protected from spilling into surface waters and sewers.

7.3 Main Control Measures Available to Reduce Contamination of Water Sources

> **TOPIC FOCUS**
>
> **COD, BOD and TOD**
>
> - **Chemical Oxygen Demand (COD)**
>   - The COD test measures materials in a water sample that can be chemically oxidised.
>   - The test is performed in a laboratory by reacting the water sample with a strong chemical oxidising agent, such as potassium dichromate, for a specified time (usually one or two hours) at a defined temperature.
>   - In essence, a COD test determines the amount of organic matter by measuring the amount of oxygen the sample will react with.
>   - COD is expressed in milligrams of oxygen per litre (mg/l).
>   - The COD test is relatively simple and can be performed within about two hours.
> - **Biological Oxygen Demand (BOD)**
>   - Also known as Biochemical Oxygen Demand.
>   - The BOD test measures all the materials in a water sample that can be broken down by the action of microbes.
>   - The test is performed in a laboratory by incubating the water sample with a culture of micro-organisms for a specified time (usually five days) under defined conditions and then comparing the level of dissolved oxygen in the sample at the beginning and end of the test.
>   - BOD is expressed in milligrams of oxygen per litre (mg/l).
>   - The BOD test is more complex and time-consuming than the COD test.
> - **Total Oxygen Demand (TOD)**
>   - Measures all of the organic and inorganic compounds present in a sample of water that can be oxidised.
>   - TOD is expressed in milligrams of oxygen per litre (mg/l).
>   - May be undertaken by online equipment that only needs a few minutes to measure a sample.
>
> **Note**: Although COD, BOD and TOD all provide measures of the potential oxygen depletion that can be caused by a polluting effluent, they do not give exactly comparable results. For example, the COD test does not measure the oxygen-consuming potential of certain organic compounds such as acetate, whereas acetate can be metabolised by micro-organisms and would therefore be detected by the BOD test. On the other hand, the oxygen-consuming potential of cellulose would not be picked up by a standard BOD test but would be detected by a COD test. It is important, therefore, that the most appropriate test is used in each situation.

Controls for Storage and Spillage

Preventing Spillages

Since spillages of noxious chemicals are a ready source of pollution, the most effective strategy is to prevent spills in the first place:

- Sloppy chemical transfer practices create an unnecessary risk of spillage, whereas more careful operating procedures prevent or minimise such losses.
- Maintenance and inspection will identify potential or actual spills and leaks early on, preventing them from either developing or getting worse. For example, corrosion, if allowed to develop unchecked, will ultimately cause the container/pipe to fail.

Corrosion of containers will ultimately cause them to fail

Main Control Measures Available to Reduce Contamination of Water Sources — 7.3

- Proper storage of materials will also help prevent spillage, e.g. siting dangerous chemicals away from internal traffic routes or with barriers to protect from collision.

> **MORE...**
>
> For further information on spillage control management, see the publication GPP 22 *Dealing with spills*, available at:
>
> www.netregs.org.uk/media/1643/gpp-22-dealing-with-spills.pdf

Wastewater Treatment Lagoons

Lagoons can be used as a wastewater treatment technique. They can be of the following types:

- Wastewater treatment, storage and evaporation lagoons, e.g. for sewage treatment, food manufacturing and agricultural processing.
- Sedimentation basins and leachate collection ponds for landfill.
- Irrigation dams used for holding and treating wastewaters.
- Ponds used for collecting potentially contaminated stormwater run-off from sites.
- Processing and wastewater lagoons used in mining, wastewater treatment and manufacturing.

If lagoons are not controlled appropriately they can present a significant risk to the environment, leading to:

- Surface water pollution (e.g. through breach of lagoon walls).
- Groundwater pollution (e.g. through the use of an inappropriate liner).
- Odour and health impacts.

Keeping Systems Separate

Appropriate Storage of Incompatible Materials

When incompatible materials come into contact with each other, e.g. during an accidental spill, the substances may react together to cause a fire or explosion or to form a toxic substance. Careful consideration of storage requirements given in safety data sheets will assist in determining the appropriate storage arrangements. Consider the following:

- Are materials likely to result in a violent chemical reaction if they come into contact with one another?
- Should a fire occur involving one material, would fire-suppression substances, such as water, cause a problem with other materials?
- Are flammable goods stored away from oxidising agents?
- Would a spillage of one material damage or disintegrate the packaging and containers of other stored materials?

Bunding of Chemical and Oil Stores

While specific legislation in a country or region should be consulted on the storage of chemicals and oils, one key requirement is to use a suitably designed and constructed bund.

Storage of hazardous liquid materials in drums and Intermediate Bulk Containers (IBCs) also needs secondary containment. The recommended size of the secondary containment volume for oil storage in England and Wales is identified in the following table; it also provides a good guide to the secondary containment volumes required for other liquid substances:

7.3 Main Control Measures Available to Reduce Contamination of Water Sources

> **DEFINITIONS**
>
> **BUND**
>
> A secondary, impermeable container in which the primary container sits. Commonly used for larger storage vessels, bunds typically consist of a wall surrounding the primary container, the inside surfaces (and floor) all being rendered impermeable. The bund is sized to 110% of the volume of the primary container.
>
> **DRIP TRAY**
>
> A simple tray placed under storage containers to collect minor leaks and spills.
>
> **INTERMEDIATE BULK CONTAINER (IBC)**
>
> A container used for the storage and transport of liquids and other bulk materials.
>
> They are cubic in shape and usually constructed of plastic surrounded by a metal cage. They often have pallet bases so they can be moved by a forklift truck.

Container Type	Minimum Secondary Containment Volume
Single drum	Secondary containment for drum storage can be provided by a drip tray with at least 25% of the volume of the drum.
Multiple drums	Secondary containment for drum storage can be provided by a drip tray with at least 25% of the total drum storage.
Single IBC	Secondary containment with at least 110% of the container volume. (You can't use a drip tray with only 25% storage capacity if you're storing oils in an IBC.
Multiple IBCs	Secondary containment with a minimum of either 25% of the total volume of the containers, or 110% of the largest container, whichever is the greater volume.

Source: Based on GPP 26 *Safe storage of drums and intermediate bulk containers* (IBCs), SEPA, NRW and NIEA, 2018 (https://www.netregs.org.uk/media/1681/gpp-26-safe-storage-of-drums-and-ibcs.pdf)

Spillage Control Management

When a pollution incident occurs, the spillage can escape from a site via different routes, including:

- Through surface water drainage system.
- Directly into a watercourse.
- Through soil, soakaways, damaged drains and surfaces to groundwater.
- To the foul sewer.

The methodology for spillage control management is:

- **Pollution Risk Assessment**

    A risk assessment should take into account all the above routes, in addition to issues such as:

    - The properties (physical, chemical and biological) of the pollutants spilt.
    - Impacts of accidents.
    - Vandalism.
    - Containment failure.
    - Flood risk.

Main Control Measures Available to Reduce Contamination of Water Sources — 7.3

- **Pollution Incident Control Plan**

  Following determination of the risk of pollution, a pollution incident control plan can be developed to ensure that an effective response is in place, should an incident occur.

  Training of staff is important to ensure that the plan is effectively implemented. For example, staff should:

  - Know what they should and should not do following a spill.
  - Be aware of where pollution control equipment and PPE are located and the location of the pollution incident control plan.

  When planning a spill response, the following hierarchy should be considered:

  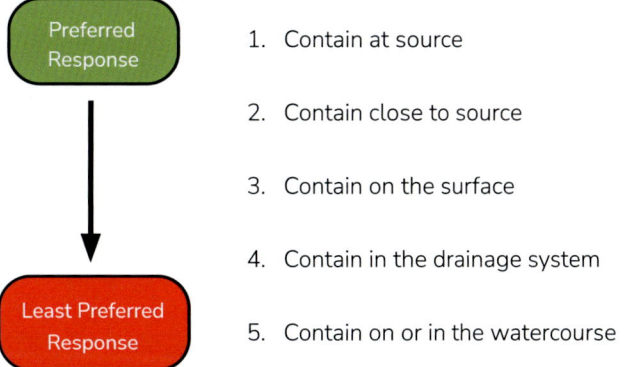

  The pollution control hierarchy

  - **Contain at source:** the most effective measures are to stop a spill at the source where the primary or secondary containers have been breached, such as sealing the damaged container/pipework with proprietary sealant or turning the container so the damaged area is at the top.
  - **Contain close to source:** this includes moving the leaking material to an undamaged container, using sorbent product to soak up the spill or using a small container to capture the spill.
  - **Contain on the surface:** the next option is to prevent the material from entering the drainage system or unsurfaced ground. Methods to achieve this include booms and drain mats.
  - **Contain in the drainage system:** if a spill has entered the drainage system then it should be retained there and prevented from entering the environment. This can be achieved by shutting valves, blocking drains or closing oil separators (interceptors - see below).
  - **Contain on or in the watercourse:** environmental damage can be reduced by containment on the watercourse prior to the spill spreading. This can be achieved by deploying a boom or damming a watercourse.

- **Site-Specific Pollution Control Options**

  The pollution risk assessment may identify that site-specific pollution control will be needed, such as on-site structures that can be used to divert or pump a spill to provide pollution containment. Examples include containment lagoons and ponds, tanks, sacrificial areas and pits and trenches.

- **Spill Clean-Up**

  Any spillage needs to be cleaned up and disposed of in line with legal requirements for waste. A review should also be completed of how the incident occurred and the effectiveness of the response plan.

7.3 Main Control Measures Available to Reduce Contamination of Water Sources

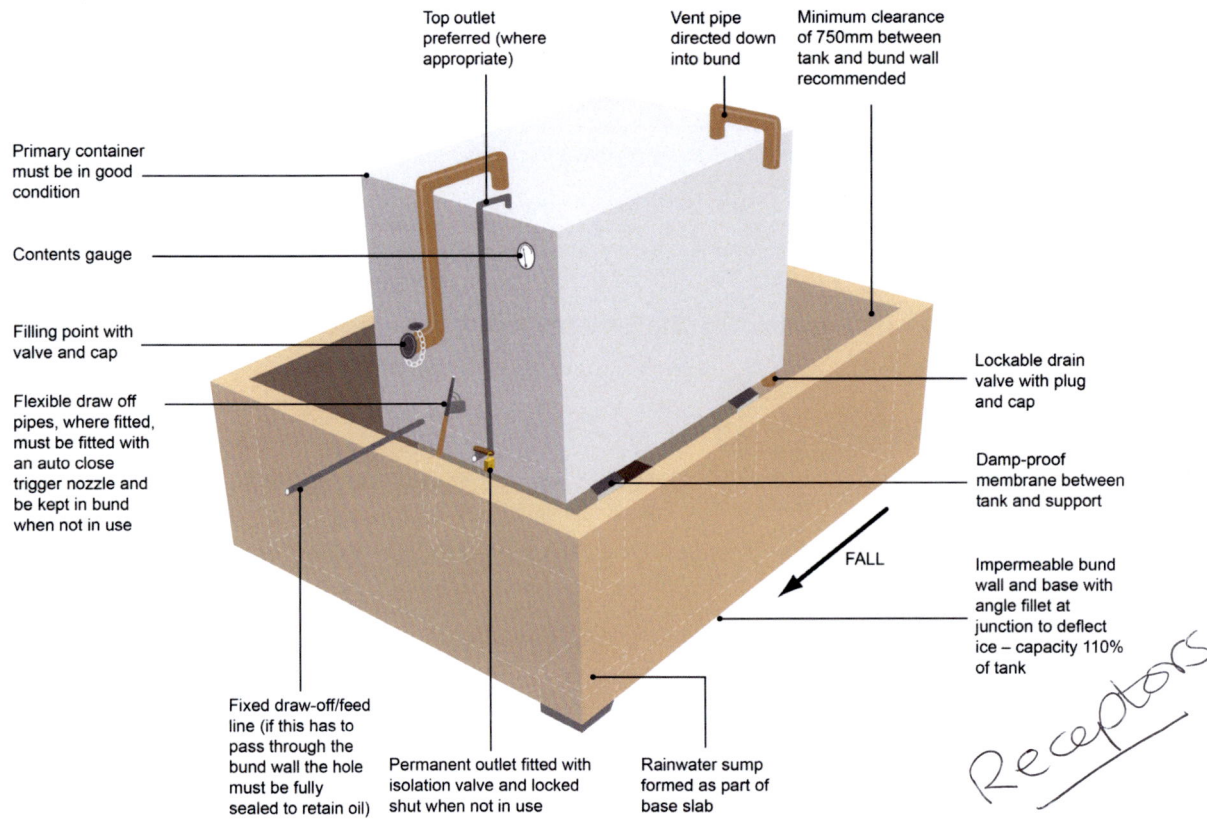

Single-skinned oil tank within an open bund (based on PPG2 *Above ground oil storage tanks*, Environment Agency, SEPA and NIEA, 2011)

Note: PPG2 has been withdrawn but remains technically correct.

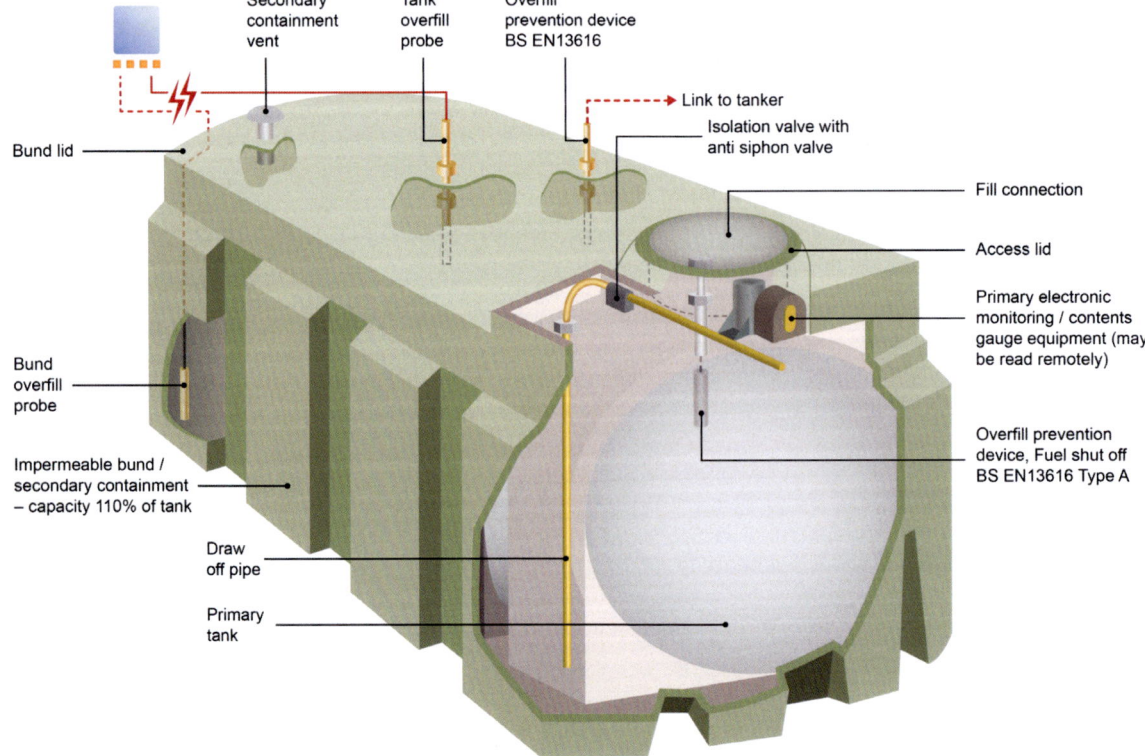

Integrally bunded tank (based on PPG2 *Above ground oil storage tanks*, Environment Agency, SEPA and NIEA, 2011)

Main Control Measures Available to Reduce Contamination of Water Sources — 7.3

Use of Oil Interceptors

Oil interceptors use the fact that oil (including oil-based fuels) floats to prevent it being discharged. Regular inspection of interceptors is essential to ensure they are not blocked or overloaded with excess volumes of oil. Different types of oil interceptors are available for different uses. For example, oil interceptors are used in surface water drainage systems from hard standings such as car parks (where obviously oil leaks from car engines can build up).

Decanter Centrifuge

A centrifuge operates to clarify an effluent around a centre line. The unit rotates at high speed around its centre line and, as it does so, the impact of gravity is replaced by that of centrifugal forces which can be around 4,000 times that of gravity. Such force can be used to cause effective separation of solids from liquids at a much faster rate than sedimentation.

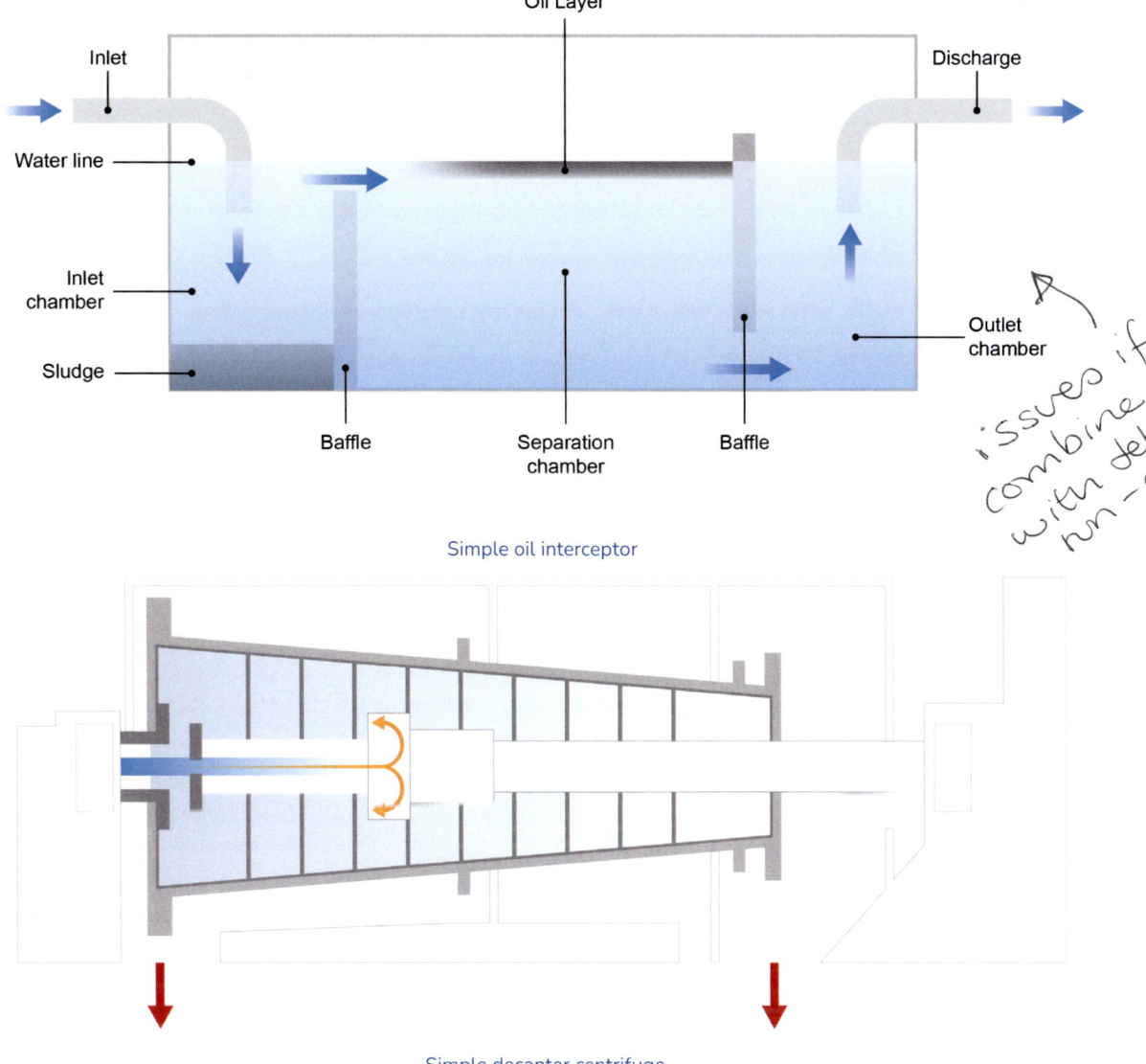

Simple oil interceptor

Simple decanter centrifuge

7.3 Main Control Measures Available to Reduce Contamination of Water Sources

Separation and Marking of Drain Systems

Sewerage and surface water systems should not mix. Process water should also be kept separate, if possible, as this will enable any sources of pollution to be more easily identified. Drain covers should be marked with both the type of drain (surface, sewer, process, etc.) and the direction of flow. A clear colour-coding system should be used and the direction of flow should not be marked on the cover itself but on the surround. Usually, blue is used to denote surface water drainage (i.e. for uncontaminated rainwater) and red for foul water drainage (i.e. for sewage and/or trade effluent).

Dealing with Spillages

Provision of spill kits suitable to deal with the type of pollution likely to occur, and training in the proper use of the kits, are an important control system. The kits must be maintained and available at the locations where spills are likely to occur, as quick action is required if pollutants are to be prevented from entering the water source.

Controls for Wastewater

Screening

This is a simple process which uses a screen (e.g. stainless steel mesh) to filter out large solids and organic matter (such as sticks, weeds), commonly used in water treatment works.

Solids Separation and Removal of Organic Load (Coagulation)

Fine particles, such as clay, metal oxides and some organic substances are difficult to settle out of suspension under natural conditions. Coagulants are used to encourage these particles to come together in what are known as 'flocs'. Aluminium is a commonly used coagulant. Once the coagulants have been added, the water must be mixed at high speed to ensure effective mixing takes place. Once thoroughly mixed, the water is passed to another tank where it is stirred slowly, allowing even larger flocs to form. Eventually, the water moves to another tank where there is very little movement and the flocs sediment out to the bottom.

Sedimentation/Flotation

Sedimentation is where the water is stored in a tank and any suspended solids are able to sink to the bottom under gravity. Alternatively, flotation can be used, where air is blown into the water, increasing the buoyancy of the particles as they absorb air. When they reach the surface, they can be skimmed off using rotating blades.

Filtration

Filtration is a separation technique whereby solids are trapped in a filter medium and the liquid is allowed to pass through. Depending on the nature and extent of the solids loading, different media can be used. For example, tertiary treatment of water in a sewage works would typically involve the use of a sand filter (anthracite may also be used).

Primary treatment of sewage commonly uses **biological** or **trickling filters**. This is where primary settled sewage is intermittently spread by a rotating distributor tube over a bed of gravel. Liquor flows over the surface of the gravel, on which a biofilm of micro-organisms develops and grows by digesting the sewage. It seeps down and is collected at the bottom. It is important that the beds do not become waterlogged.

Main Control Measures Available to Reduce Contamination of Water Sources 7.3

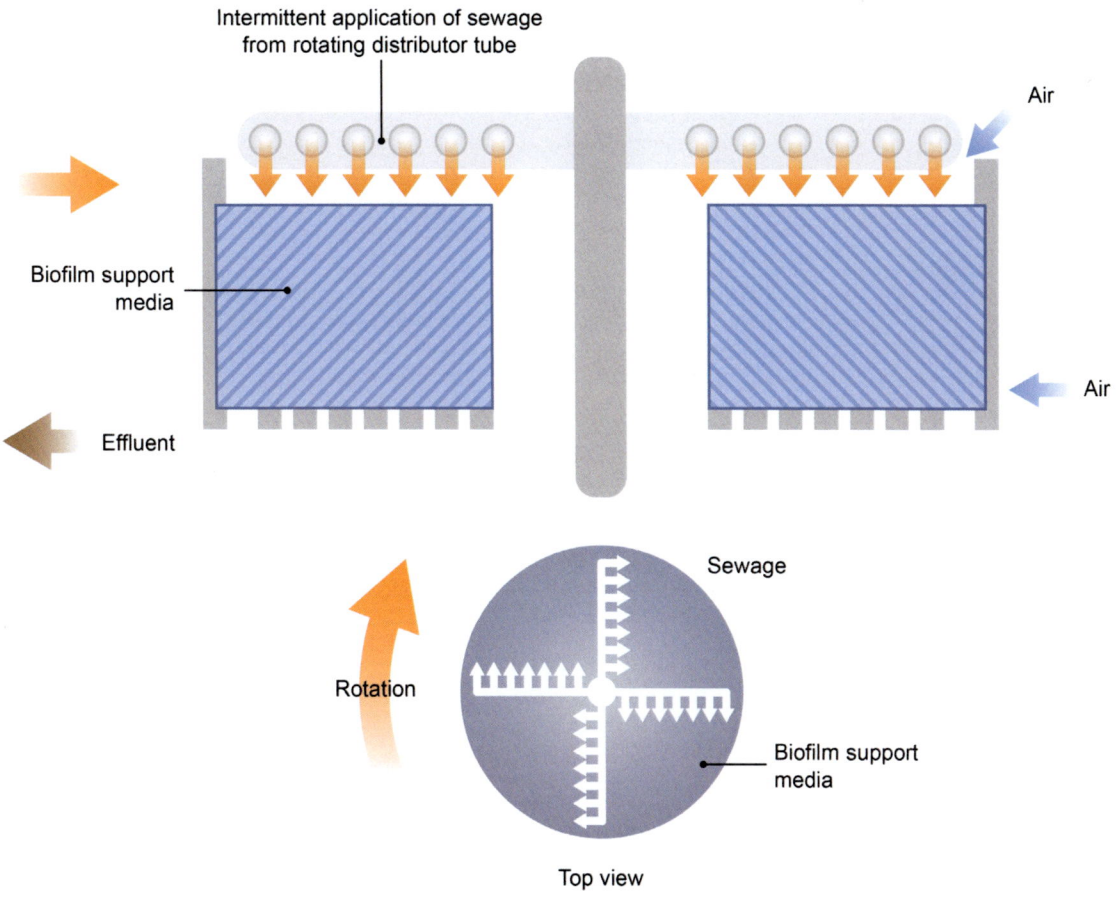

Trickling filter system - cross-section and top view

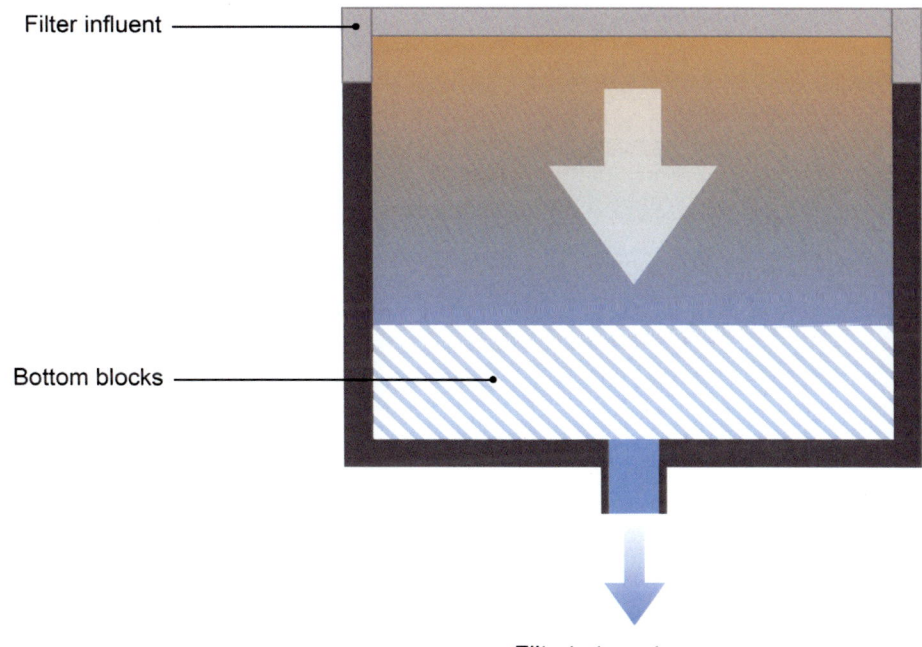

Sand filter filtration cycle

7.3 Main Control Measures Available to Reduce Contamination of Water Sources

Centrifugal Separation

Centrifugal separation is really a form of accelerated settling. Normal settling leads to relatively slow separation of solids from liquids, forming a sediment at the bottom under the influence of gravity. In centrifugal separation, the water is fed into a centrifuge that spins at high speed. The centrifugal forces act on the heavier particles in the water, forcing them to the outside where they are collected and fed away from the water. The clean water passes through the system. The technique is typically used to de-water sludge (from sewage treatment operations).

Correction of pH

As mentioned earlier, for discharge consents and permits, the pH has to be adjusted to within certain limits. If the wastewater is too acidic, it can be adjusted with alkaline materials such as lime (calcium oxide/hydroxide) or sodium carbonate. If it is too alkaline, it can be adjusted with acids such as hydrochloric acid.

Difficulties in Maintaining Equipment in Some Locations or Environments

As with other types of pollution control equipment, any equipment that is used to control water pollution must be subject to a planned preventive maintenance programme. It is important, therefore, that funding for maintenance is provided to ensure the devices' continuing effectiveness.

Poor maintenance of pollution abatement and other equipment is a key reason for process failure and pollution incidents, often as a result of a lack of proper funding. It should also be recognised that certain environments present significant logistical or practical challenges for planned preventive maintenance.

> **STUDY QUESTION**
>
> 4. List five methods used to reduce contamination of water resources.
>
> (Suggested Answer is at the end.)

Summary

This element has dealt with the control of contamination of water sources.

In particular, this element has:

- Explained that water supply companies have a legal duty to supply water that is fit to drink, sourced from groundwater, reservoirs and rivers. Varying levels of purification are required to produce water which is clear, palatable, safe and reasonably soft.
- Outlined the water cycle and sources of water.
- Emphasised that water should be treated as a valuable resource. Methods of water conservation include dual flush toilets, installation of a water meter and grey water recycling.
- Outlined the main sources of water pollution, including domestic wastewaters, surface-water drainage, industrial discharges, process and cooling water, drainage from mining, agricultural operations, contamination from spills and leaks, solids such as grit, plastics, etc.
- Outlined the main control measures available to reduce contamination of water sources, including:
  - Permits for discharge to surface, groundwater and public sewer systems.
  - Controls for storage and spillage: prevention of spillages in the first place with the use of appropriate procedures and techniques, appropriate storage, wastewater lagoons, separation and marking of drain systems, use of oil interceptors, bunding of chemical and oil stores and clearing of spillages.
  - Controls for wastewater: screening, solids separation and removal of organic load (use of 'flocs'), centrifugal separation (accelerated settling), sedimentation/flotation, filtration (solids are trapped in a filter medium and the liquid passes through), and correction of pH.

Exam Skills

Question

Scenario
During a routine inspection of a transport yard, a manager has observed oil floating on the surface of a small stream that runs alongside the yard. The oil is immediately downstream of a surface water discharge drain in the transport yard.

Task
What are the checks that the manager should make in the initial investigation of this incident? **(8 marks)**

Approaching the Question

Think about the steps you would take to answer the question:

- Read the scenario carefully. With this question you need to develop some ideas on the checks that should be undertaken when first coming across the incident. Note: the question is not asking to control the oil pollution or reduce the risk of it occurring.
- Now look at the task – prepare notes on:
- What tasks are occurring in the transport yard that could lead to leaks of oil.
- How the oil could travel from these sources to the stream.
- Consider the marks available. In this case, there are 8 marks available so you should provide 8 items to check for oil.
- Read the scenario and task again to make sure you understand them and have a clear understanding of the sources of oil in the transport yard and what to check.
- Jot down an outline plan - this might include:
    - Vehicles, bulk storage tanks, drum storage, machinery, yard surface, oil/water separators, drainage inlets, drainage system.

Now have a go at the question yourself.

Exam Skills

Example of How the Question Could be Answered

The checks that should be made in the initial investigation of this incident include consideration of:

- Vehicles in the yard, to determine if an oil leakage has come from a poorly maintained vehicle.
- Bulk storage tanks in the yard, to determine whether primary containment (the tank itself) and, if present, secondary containment (a bund) has been breached.
- Areas of the yard where drums are being stored; those not being stored in secondary containment present a greater risk.
- The condition of any machinery that is being stored in the yard, which may be the source of the leak.
- The presence of any spillages on the yard surface or residue from spillages that have previously occurred.
- Oil/water separators as these may not have been maintained correctly and oil may be overflowing into a surface water drain.
- Drainage inlets, as oil residue may be present at inlets into the surface water drains.
- The drainage system on the site, to check whether any oil from the site has entered the site's surface water drains.

Element 8

Control of Waste and Land Use

Learning Outcomes

- Understand the issues associated with waste and support responsible waste management.

Learning Objectives

Once you've read this element, you'll be able to:

1. Demonstrate awareness of common waste types, the outlets available for waste, and environmental issues associated with waste and contaminated land.

2. Suggest suitable waste management measures, applying the waste hierarchy.

Contents

Waste Types	**8-3**
The Waste Framework Directive	8-3
Definition of Waste	8-3
Inert Waste	8-4
Hazardous Waste	8-4
Non-Hazardous Waste	8-5
Clinical Waste	8-5
Radioactive Waste	8-6
Waste Types Subject to Specific Legal Requirements	8-6
Types of Waste	8-6
Minimising Waste	**8-8**
Impacts from Waste	8-8
Waste - a Worldwide Problem	8-8
The Business Case for Minimising Waste	8-9
The Waste Hierarchy	8-10
Applying the Waste Hierarchy at Every Stage	8-11
Managing Waste	**8-13**
What is the Waste Chain?	8-13
Recognition of the Key Steps	8-13
Responsible Waste Management	8-13
Benefits, Limitations and Barriers to Re-Use and Recycling	8-14
On-Site Separation and Storage including Segregation, Identification and Labelling	8-15
Transportation including Transfer to an Authorised Person and Required Regulatory Documentation	8-16
Differing Legal Requirements for Waste	8-17
Disposal	8-18
Producer Responsibility	8-18
Packaging Waste	8-18
Electrical and Electronic Waste	8-18
Waste from Construction Projects	8-20
Outlets Available for Waste	**8-21**
Circular Economy	8-21
Landfill and Incineration as Ultimate Disposal Routes	8-22
Other Treatment or Disposal Routes	8-25
Global Waste Trade	8-26
Export Costs and the Impact of Export, Landfill and Aggregate Taxes	8-26
Risks Associated with Contaminated Land	**8-28**
The Potential Effects of Contaminated Land on the Environment	8-28
Contaminated Land Liabilities	8-29
Summary	**8-31**
Exam Skills	**8-32**

Waste Types

IN THIS SECTION...

- Waste can be defined as "*any substance or object which the producer or the person in possession of it discards, or intends or is required to discard*".
- Waste can be categorised in a number of ways. Common categories are: inert, hazardous, non-hazardous, clinical and radioactive.
- Types of waste include general, municipal, electronic, organic, construction/demolition, industrial and agricultural.

The Waste Framework Directive

'Waste' is a term we all think we understand, but the legal issues surrounding the definition of waste are complex. There are many different types of waste that may be encountered in practice, and it is important to understand the various categories because of the differences in the way that they are regulated.

The **EU Waste Framework Directive**, first adopted in 1975, is the foundation of waste regulation and aims to ensure a uniform approach to waste management across the EU. Member states must:

- Adopt the waste hierarchy (i.e. give priority to waste prevention and encourage re-use and recycling).
- Ensure that waste is handled safely and without harming the environment.
- Ensure that waste management activities are authorised.
- Establish an adequate infrastructure of waste management installations.
- Prepare waste management plans.
- Ensure that waste producers bear the costs of disposal in line with the 'polluter pays' principle.

Inert waste

Definition of Waste

Article 3(1) of the current version of the Directive (**2008/98/EC**) retains the original 1975 general definition of waste as:

> "*...any substance or object which the producer or the person in possession of it discards, or intends or is required to discard*".

Anything discarded or dealt with as waste must be presumed to be waste unless proved otherwise. A 'Yes' answer to any of the following questions should clarify any doubts about the matter:

- Would it normally be described as waste?
- Is it a scrap material?
- Is it an effluent or other unwanted substance?
- Is it broken, worn out, contaminated or spoilt?
- Is it being discarded as if it were waste?

Waste materials can be categorised in a number of ways. The categories that are most typically recognised in legislation around the world are described below.

8.1 Waste Types

> **MORE...**
> Further information on the legal definition of waste can be found at:
>
> www.gov.uk/government/publications/legal-definition-of-waste-guidance

Inert Waste

Broadly speaking, this is waste which is stable, i.e. it does not degrade physically, chemically or biologically, nor does it dissolve, burn, chemically react or leach out to any degree that could be considered ecotoxic. Examples would include uncontaminated bricks, glass, concrete and tiles. If there is any suspicion of contamination, these items cannot be considered inert waste. The EU **Landfill Directive 99/31/EC** provides further criteria on what constitutes inert waste.

> **DEFINITION**
>
> **ECOTOXIC**
>
> Generally taken to mean 'damaging to the environment', although it is a general term and does not account for levels of toxicity, e.g. very toxic or toxic. Nor does it account for the sensitivity of specific species or ecosystems, e.g. an ecotoxic substance may be very toxic to one species but have little, if any, harmful effect on another.

Hazardous Waste

Certain wastes pose a particular danger to human health or to the environment. The **Waste Framework Directive** (Annex III) identifies the properties of a waste material that render it hazardous. In summary, this covers substances that are:

- Explosive.
- Oxidising (substances which are highly reactive in contact with other substances).
- Highly flammable.
- Irritant (substances that can cause inflammation of the skin or mucous membranes).
- Harmful (substances which may involve limited health risks).
- Toxic (substances that may involve serious health risks).
- Carcinogenic (substances that may induce cancer).
- Corrosive (substances that may destroy living tissue on contact).
- Infectious (substances that contain disease-causing micro-organisms).
- Mutagenic (substances that may damage or change hereditary genetic material).
- Waste which releases toxic gases in contact with water, air or an acid.
- Sensitising (substances that can elicit an allergic reaction).
- Ecotoxic (substances that present a risk to the environment).
- Waste that may yield another substance after disposal that exhibits any of the above properties (e.g. waste in a landfill that generates a toxic leachate).

Examples of some commonly encountered wastes that meet one or more of these hazardous waste criteria are:

- Liquid fuels, such as petrol and diesel; solvents such as white spirit (explosive; highly flammable).
- Strong acids or alkalis, e.g. battery acid or bleach (oxidising; irritant; corrosive).
- Insecticides, wood preservatives or old medicines (harmful; toxic; ecotoxic; sensitising).
- Waste oil, batteries containing lead, cadmium or mercury, fluorescent lighting tubes containing mercury (toxic; ecotoxic; mutagenic).
- Contaminated textiles, such as used bandages or dressings, asbestos (infectious; carcinogenic).

Disposal of hazardous wastes is managed by specialist companies who operate dedicated chemical plants. Recovery of waste oils, solvents, etc. is usually followed by incineration of the residues, which must be carefully controlled to minimise production of substances such as dioxins and furans. This involves careful control of incinerator temperatures and cleaning of effluent gases.

(Note that in Scotland, the term 'special waste' is equivalent to 'hazardous waste' in most other countries.)

Non-Hazardous Waste

Wastes which are controlled under legislation, but are neither inert, nor exhibit any of the properties of hazardous wastes, are classified as non-hazardous. This category actually accounts for a high proportion of the wastes that are generated on a day-to-day basis by households and businesses, including paper, card, plastic packaging, cans and food waste.

Despite the name, non-hazardous wastes have the potential to cause significant environmental impacts. Many of these wastes are biodegradable, or may be corroded by the action of weather. If these wastes are landfilled, they may generate:

- Methane gas, which is a potent greenhouse gas.
- Toxic leachate with the potential to contaminate surface and groundwaters.

Clinical Waste

Clinical waste is, effectively, a special category of hazardous waste, often treated separately in legislation because of the need for special methods of treatment and disposal. Clinical wastes are healthcare wastes which could harm people if they come into contact with them. The definition is wide-ranging, but includes:

- Soiled surgical swabs, dressings, etc.
- Excretions.
- Blood or body fluids.
- Human and animal tissues, carcasses, etc.
- Syringes, needles or other sharps.
- Drugs or other pharmaceuticals.

Clinical wastes and healthcare wastes

Clinical waste should be **segregated** from general waste; **separate bins, signage and training** should be provided to encourage this. There are various methods for achieving this, for example:

- Soiled surgical dressings should be put into heavy-duty yellow bags (2/3 full) and securely fastened.
- Sharps should go into properly designed sharps containers.
- Laboratory material, where risk of pathogens is high, should be autoclaved before being included with other clinical waste.

8.1 Waste Types

> **DEFINITION**
>
> **AUTOCLAVE**
>
> Equipment which uses high-pressure steam to sterilise material.

Radioactive Waste

Radioactive waste is also a special category of hazardous waste that is governed by specific legislation. The International Atomic Energy Agency (IAEA) defines radioactive wastes as being:

"waste that contains or is contaminated with radionuclides at concentrations or activities greater than clearance levels as established by the regulatory body".

In practice, this definition covers:

- High volumes of waste from the nuclear power industry where the level of radioactivity may vary from low to very high.
- Low volumes of waste produced by other businesses that use small quantities of radioactive materials in laboratories, and in sensing and monitoring equipment.

Radioactive warning sign

Waste Types Subject to Specific Legal Requirements

Some waste types are subject to specific legal requirements. In the UK, for example, the term 'controlled waste' is often used. Controlled waste is any waste which is covered by specific parts of the **Environmental Protection Act 1990**.

Controlled waste effectively covers all of the waste that is likely to be encountered, including that from households, commerce (including construction and agriculture) and industry (including mining and quarrying).

Types of Waste

- **General Waste**

    This is waste materials from businesses and households that cannot be recycled. It incorporates many waste types including non-recyclable plastics, metal-containing items and packing materials such as polystyrene. Usually, general waste is disposed of to landfill, but is often segregated first to remove recyclable materials.

- **Municipal Waste**

    This is a term that covers waste collected from households usually by a local authority. This waste type may consist of various items, such as product packaging, garden waste, glass, wood and plastics.

- **Electronic Waste**

    This includes electronic equipment and all components, subassemblies and consumables which are part of the product at the time of discarding. Some elements of electronic waste are often hazardous, containing harmful metals such as cadmium or brominated flame retardants.

- **Organic Waste**

    This is waste materials that are biodegradable. They can be broken down and produce carbon dioxide, methane or simple organic molecules which can pose significant environmental issues when, for example, they are disposed of to landfill. Examples of organic waste include garden waste, food waste and wood.

Waste Types | 8.1

- **Construction and Demolition Waste**

    Waste materials from construction and demolition sites can vary but they may include materials such as metal, glass, bricks and plastic packaging in addition to hazardous wastes such as adhesives or sealants. It is common for construction and demolition organisations to develop site waste management plans to provide a management framework for such wastes.

- **Industrial Waste**

    This incorporates any waste that is produced from industrial activity during a manufacturing process. It can include many types of waste such as chemical waste, metals, oils and paper products.

- **Agricultural Waste**

    Agricultural waste is any substance or object that has been used in agriculture or horticulture that is discarded or is intended to be discarded. Again, it could incorporate numerous waste types but examples may include empty pesticide containers, used tyres or old silage wrap.

STUDY QUESTIONS

1. Define 'waste'.
2. Identify the criteria used to classify waste as hazardous.

(Suggested Answers are at the end.)

8.2 Minimising Waste

Minimising Waste

IN THIS SECTION...
- Minimising waste can help organisations make significant cost savings by following some practical steps, such as reviewing current practice, identifying opportunities for improvement, setting KPIs and targets, training and monitoring performance.
- There are many business benefits from minimising waste.
- The waste hierarchy defines the most desirable methods for waste management through to the least desirable.

Impacts from Waste

Activities involving waste can be a source of many pollution issues both direct and indirect. Poor waste storage, for example, may lead to water pollution and land contamination from spillage of liquid waste; and dusty waste materials may be released during transfer, causing air pollution.

Waste - a Worldwide Problem

The increasing volume of waste generated by the developed world is one of the biggest problems facing the planet today. In the developing world, it is also a major worry, with increased populations and higher standards of living, both contributing to significant increases in waste. Packaging is often highlighted as an area of significant waste and there are still major reductions that can be made in many areas of packaging. However, not all packaging is bad. Much of it protects goods from being damaged in transit and, without some packing, more damage, and therefore waste, would occur. As an example, a shrink-wrapped cucumber remains saleable for 14 days, whereas without wrapping it lasts just three days.

Waste - a worldwide problem

> **TOPIC FOCUS**
>
> **Waste Minimisation in Practice**
>
> Waste minimisation has clear benefits for society but can also result in significant cost savings for organisations. Some practical steps for achieving waste minimisation are outlined below:
>
> - **Current Practice**
>
>   The process should start by undertaking a review of current practice. Measurement is at the heart of this activity:
>
>   - How much of exactly what type of waste is currently being created in particular areas of the organisation?
>   - Waste arisings should be classified and quantified by type (e.g. paper, card, scrap metal, empty containers, chemicals, waste raw material, etc.) and by category (i.e. inert, hazardous, non-hazardous).
>   - What disposal routes are currently being used for each type of waste? Which disposal contractors are being used?
>   - How much did it cost in the last year to dispose of each waste stream?
>
>   Note that much of the information and data required can be obtained by reviewing statutory documentation (such as written information and hazardous waste consignment notes in the UK - see later).
>
>   (Continued)

Minimising Waste 8.2

> **TOPIC FOCUS**
>
> - **Identify Opportunities**
>
>   Having established what is happening at the moment, identify potential improvements:
>
>   - Is the organisation ordering surplus goods that it does not need?
>   - Are manufacturing processes using raw materials efficiently?
>   - Are there opportunities for investing in more resource-efficient equipment or processes?
>   - Can any materials that are currently going to waste be re-used elsewhere in the organisation?
>   - Is it possible to switch to more efficient waste contractors?
>
> - **Set KPIs and Targets**
>
>   It should now be clear what the current unit costs for waste disposal are and what improvements can reasonably be achieved. This information can be used to:
>
>   - Set KPIs (e.g. waste produced per unit of production or output; wastes costs per unit of production).
>   - Set improvement targets.
>
> - **Responsibilities and Training**
>
>   - Nothing will happen unless clear responsibilities are identified for waste management.
>   - Everybody in the organisation needs to be trained so they can play their part, e.g. understanding how particular wastes should be dealt with on site.
>
> - **Monitor Performance**
>
>   - Performance against KPIs and targets needs to be assessed.
>   - Cost savings achieved need to be calculated.
>   - Improvements need to be communicated widely within the organisation. Cost savings are powerful incentives for supporting waste minimisation programmes.

> **MORE...**
>
> WRAP was established to facilitate improvements in resource efficiency in the UK, and has produced a number of useful publications to help organisations minimise waste. These are available from:
>
> www.wrap.org.uk/category/subject/waste-reduction

The Business Case for Minimising Waste

There are many business benefits that may occur from minimising waste; these include:

- financial savings;
- reduction in raw materials use;
- improved corporate image;
- less chance of prosecution;
- improved health and safety standards; and
- increased employment opportunities.

8.2 Minimising Waste

The Waste Hierarchy

The waste hierarchy defined in the **Waste Framework Directive** is a system of applying best practice to the management of waste. The current waste hierarchy is:

The Waste Hierarchy

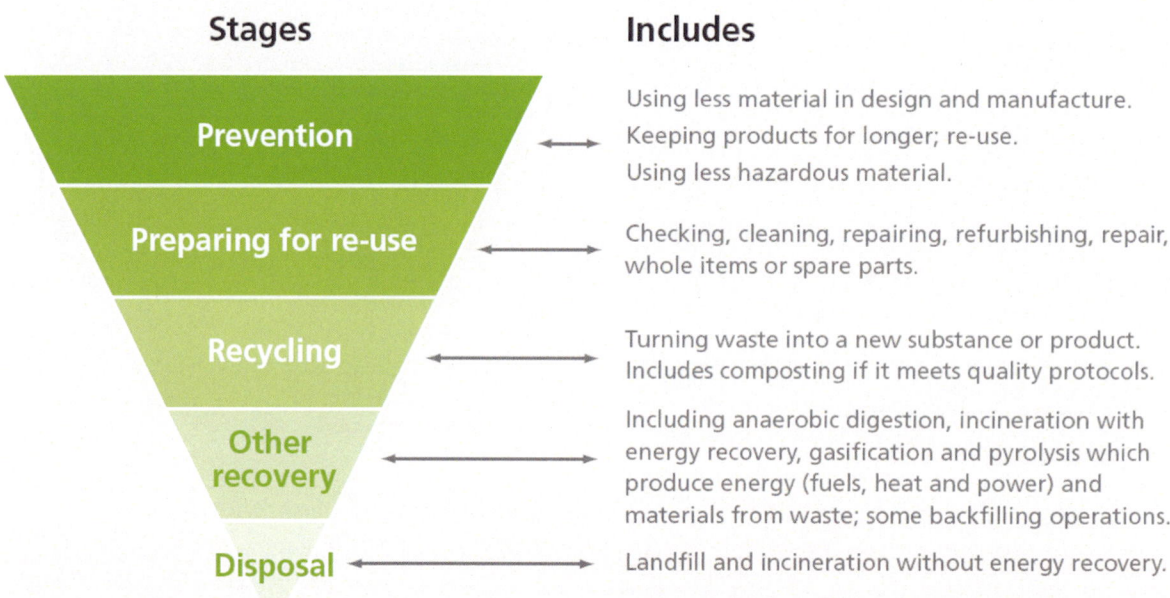

Stages	Includes
Prevention	Using less material in design and manufacture. Keeping products for longer; re-use. Using less hazardous material.
Preparing for re-use	Checking, cleaning, repairing, refurbishing, repair, whole items or spare parts.
Recycling	Turning waste into a new substance or product. Includes composting if it meets quality protocols.
Other recovery	Including anaerobic digestion, incineration with energy recovery, gasification and pyrolysis which produce energy (fuels, heat and power) and materials from waste; some backfilling operations.
Disposal	Landfill and incineration without energy recovery.

Source: Government Review of Waste Policy in England 2011, DEFRA, 2011 (www.defra.gov.uk/publications/files/pb13540-waste-policy-review110614.pdf)

Minimising Waste — 8.2

> **TOPIC FOCUS**
>
> **The Waste Hierarchy**
>
> - **Waste Prevention and Reduction**
>
>   Clearly, the best option is not to produce the waste at all. If this cannot be achieved, then producing less waste is desirable. Techniques include designing products so that they produce no or less waste in manufacture or use, or considering the repair, re-use and ability to be recycled of the full product or component part. Examples of this are soft drinks bottles that contain approximately 25% less plastic than when plastic bottles first started to be used.
>
> - **Preparing for Re-Use**
>
>   This includes activities such as repairing, cleaning and inspection so that products or components of products may be re-used without any type of pre-processing. Examples include inspecting and fixing (if required) pallets, and washing and re-using bottles for the same purpose.
>
> - **Recycling**
>
>   The range of products that are now recyclable is increasing every day. However, a product is not fully recycled until it goes back into the market and is purchased as a new product. Glass, metal and paper are all frequently recycled. In recent years, plastics have become more acceptable to recycling as new products have been developed.
>
>   Composting involves the breakdown of organic materials, such as food and garden wastes by bacteria, fungi and insects into fertiliser. There are many local authority collection and composting schemes or it can be done locally at your organisation or home.
>
> - **Other Recovery**
>
>   Energy can be recovered from waste, e.g. by recovering the heat generated when burning waste such as at waste-to-energy incinerators.
>
> - **Disposal**
>
>   Responsible disposal should only be considered if none of the above options are appropriate.
>
>   The two main methods for final disposal are landfill and incineration without energy recovery, which we will look at in more detail later.

Applying the Waste Hierarchy at Every Stage

It is important to note that the waste hierarchy should be applied at every stage of the life cycle of a product. This holistic view will ensure that the maximum quantity of waste will be minimised.

Cleaner design, for example, involves initially determining how a product impacts on the environment during its life cycle (raw materials, manufacturing, transportation, use and end of life) and then determining how these impacts could be reduced through better design. A product's impact on the environment can be established by looking at the following:

- Reduced raw material use.
- Elimination of hazardous materials.
- Reduced use of energy and water.
- Less pollution and waste.
- Increased service life.
- Greater potential for recycling.

8.2 Minimising Waste

MORE...
blog.rrc.co.uk/2019/12/23/the-importance-of-good-design-to-reduce-environmental-impact/

STUDY QUESTION

3. List the five basic elements of the waste hierarchy.

(Suggested Answer is at the end.)

Managing Waste

IN THIS SECTION...

- Responsible waste management requires consideration of:
    - Segregation, storage and labelling.
    - Transport methods and documentation required for transport.
- Differing requirements for waste management exist in different countries; managing waste in a particular country requires a good knowledge of that country's environmental legislation.
- Specific EU Directives exist which set out requirements relating to:
    - Packaging waste.
    - Electrical and electronic waste.
- Construction sites produce significant quantities of waste - site waste management plans are often produced for large projects.

What is the Waste Chain?

The waste chain is a series of steps that waste will go through, from producer to point of final disposal or recycling, when it becomes another product ready for use, rather like a chain of custody for evidence. A,s we shall see below **Directive 2008/98/EC** on waste (referred to as the **Waste Framework Directive**) requires a producer to ensure that waste is handled correctly at each stage in the chain. Until the waste has reached a point of final disposal or been recycled, it remains the responsibility of the producer.

Waste needs to be disposed of correctly

Recognition of the Key Steps

The key steps are:

- On-site separation.
- Storage.
- Transportation.
- Disposal.

We will look at these and other aspects in more detail later, but first we will consider the issue of responsible waste management.

Responsible Waste Management

The principal legislative instrument covering waste management in the European Union is the **Waste Framework Directive (2008/98/EC)**. This is retained legislation and the UK may divert from its requirements in the future.

The Directive:

- Applies to all waste, except: radioactive wastes, waste from extraction and prospecting of mineral resources, non-dangerous agricultural wastes, wastewaters, decommissioned explosives, and unexcavated contaminated land.
- Encourages the implementation of the following hierarchy:
    - Prevention or reduction of waste production and its harmfulness.
    - Recovery of waste by means of recycling, re-use or reclamation or the use of waste as a source of energy.

8.3 Managing Waste

- Ensures that waste is recovered or disposed of without endangering human health or the environment.
- Establishes an integrated and adequate network of disposal installations.
- Requires:
    - Member states to establish or designate the competent authority or authorities to be responsible for implementing the Directive; the competent authority is required to draw up a waste management plan.
    - Undertakings which carry out waste operations to obtain a permit from the competent authority.
    - Establishments or undertakings which collect or transport waste on a professional basis to be registered with competent authorities.
    - That in accordance with the 'polluter pays' principle, the cost of disposing of waste must be borne by the holder or producer.
- Sets new recycling targets to be achieved by EU member states by 2020, including recycling rates of 50% for household and similar wastes and 70% for construction and demolition waste.
- Strengthens provisions on waste prevention through an obligation for member states to develop national waste prevention programmes and a commitment from the EC to report on prevention and set waste prevention objectives.
- Sets a clear, five-step 'hierarchy' of waste management options (as covered earlier) according to which prevention is the preferred option, followed by re-use, recycling and other forms of recovery - with safe disposal as the last recourse.
- Clarifies a number of important definitions, such as recycling, recovery and waste itself. In particular, it draws a line between waste and by-products and defines when waste has been recovered enough - through recycling or other treatment - to cease being waste.
- Establishes that member states must implement legal arrangements to ensure that:
    - Holders of hazardous waste do not mix different categories.
    - Holders of waste keep records.
    - Movements of hazardous waste are accompanied with manifests.

Note: the requirements of the Directive are covered in the UK by the **Environmental Protection Act 1990** Part II and secondary legislation made from the Act.

Benefits, Limitations and Barriers to Re-Use and Recycling

Benefits

To ensure that recycling is undertaken, the benefits of recycling should be communicated to key stakeholders. Such benefits include:

- **Raw material reduction** - if more waste is recycled then fewer raw materials need to be extracted, with a subsequent reduction in environmental impacts.
- **Corporate image** - communicating recycling schemes in corporate reports and by other means is good for an organisation's image and will help improve the reputation of the organisation, both internally and externally.
- **Pollution minimisation** - recycling results in less pollution, as extra raw materials do not have to be extracted and processed and the product is not disposed of to landfill.
- **Morale** - recycling schemes require the participation of the workforce, which should provide them with a sense of pride that the organisation is improving its environmental performance.
- **Energy reduction** - recycling saves energy and associated economic and environmental costs. Many metals require much less energy to recycle in comparison to being produced from ore. Reductions in energy will also result in minimisation of air pollutants from energy generation.
- **Cost** - recycling is generally a more cost-effective way of dealing with waste. Money can be gained for sending waste for recycling, in addition to not having to pay landfill tax.
- **Employment** - recycling creates jobs in collection and processing of wastes.

Limitations and Barriers

The barriers to recycling are many and varied, depending on the point in the waste management chain that you examine:

- **Limitations on storage space** - many homes and organisations lack space in which to segregate and store different bins for different waste streams.
- **Perceptions and attitudes** - getting individuals both at home and at work to think about recycling and separation of waste as part of their normal routine. Education is the key to overcoming this barrier. Not being rewarded for recycling can also have a negative effect.
- **Consistency of service** - differences in how local authorities collect recyclables and what they will not collect lead to confusion as to what can be recycled and what cannot.
- **Markets for recyclables** - the markets for recyclables still need significant investment to develop new processes and new products that can use recycled materials, and prices need to be closer to the 'standard' products. Often, recycled products are seen as inferior in quality and more expensive. This situation does not encourage the market to develop.
- **Legislative restrictions** - waste legislation can make it difficult to take a material out of the waste stream once it is defined as waste.
- **Inadequate encouragement for re-usable containers** - supermarkets do not encourage the use of re-usable containers and, as the majority of people do most of their shopping in supermarkets, this has a significantly adverse impact on the development of this market. There are a few small shops now opening that require customers to bring their own containers, to refill them with everything from sweets to washing powder. Supermarkets have, however, moved towards 'bags for life' rather than plastic carrier bags.

There are many other barriers to re-use and recycling, but most of these are caused by systems designed for the convenience of business. Changing these processes, often going back to good practices undertaken in the past, could bring significant positive improvements.

Additionally, the following can help overcome some barriers:

- Improving collection of waste for recycling, with more regular doorstep collections.
- Improving communications on recycling and practical advice on how to recycle.
- Ensuring that people and organisations are aware of the benefits and success of taking part in a recycling scheme.

On-Site Separation and Storage including Segregation, Identification and Labelling

Good waste management starts at the very beginning of the process. Once different waste products are mixed, it becomes increasingly difficult, and therefore expensive, to separate them later. Different wastes require different storage containers and it is important that the right container is used for the right waste. For example:

- Heavy cardboard can be stored in a container that is quite open, although it should be under cover to prevent the cardboard getting wet.
- Paper, especially shredded paper, requires a closed skip as it is easily blown away.

Waste segregation

8.3 Managing Waste

> **TOPIC FOCUS**
>
> **On-Site Separation and Storage**
>
> To ensure that waste is managed appropriately and does not escape from control, the following is required:
>
> - Prevention of:
>   - Corrosion or wear of containers.
>   - Accidental spills or leakages.
>   - Breach of containment by weather.
>   - Blowing away or falling from vehicles or storage.
>   - Scavenging by vandals, thieves, children, trespassers or animals.
> - Protection of waste while it is held (cover skips, store liquids in bunded enclosures).
> - Ensuring that waste reaches the next holder intact. (If the next stage is a waste transfer station, it will be sorted and mixed so that excessive packaging is not needed.)
> - Segregating incompatible wastes (preventing cross-contamination of waste).
> - Ensuring security (secure against waste attractive to scavengers, e.g. food waste). Waste left for collection should be adequately secured and left for a minimum of time.
> - Labelling waste where appropriate and in accordance with the hazardous substance legislation.

Transportation including Transfer to an Authorised Person and Required Regulatory Documentation

Transportation is also strictly controlled. In particular, it often requires documentation that follows the waste from the point of production to final disposal, recycling or re-use.

> **TOPIC FOCUS**
>
> **Waste Transportation**
>
> Waste holders should ensure that:
>
> - Waste is transferred only to a waste carrier, who must be registered with a competent authority.
> - Carriers are fit and suitable to handle and dispose of the waste. The holder/producer ultimately remains responsible for the fate of the waste.
>
> As we have stated previously, when transporting waste internationally, the requirements of the Basel Convention should be considered, which include:
>
> - Hazardous waste for recovery is not permitted to be exported to non-OECD countries.
> - Non-hazardous waste for recovery can be freely traded between EU member states and OECD countries. It is subject to controls stated in the regulations.
> - Hazardous waste shipped for recovery between emerging states and OECD countries must have prior written notification from the competent authority of despatch, destination and transit and their consent prior to shipment beginning.
> - Possible controls for non-OECD countries include prohibition, prior written notification and consent.

Managing Waste | 8.3

> **DEFINITION**
>
> **OECD**
>
> The Organisation for Economic Co-operation and Development is a governmental economic organisation of largely high income and developed economies such as the United Kingdom, Japan and the United States.

Regulatory Documentation

In the UK, regulatory waste documentation is required when waste is removed from a site (the systems used in other countries may be different, but share similar principles). For general waste, this takes the form of written information, which must describe the waste, the current holder and the person collecting the waste. This must be retained for two years.

In a similar manner, a consignment note is required for hazardous waste. The consignment note:

- Describes the:
    - Nature of the waste.
    - Process producing the waste.
- Includes the six-figure code from the EU List of Wastes.
- Remains with the waste until point of final disposal, being further completed by each waste carrier in the chain of custody.

Consignors and carriers of hazardous wastes must keep a register of the consignment note copies for three years; consignees of the waste must keep all such consignment note copies until they surrender the licence for the disposal site they manage.

> **DEFINITIONS**
>
> **CONSIGNOR**
>
> The person producing the waste and causing it to be removed from the premises.
>
> **CONSIGNEE**
>
> The person receiving the waste for treatment or disposal.

> **MORE...**
>
> For more information on hazardous waste, go to:
>
> www.gov.uk/dispose-hazardous-waste

Differing Legal Requirements for Waste

You should be aware that the legal requirements for managing domestic, commercial and industrial waste can differ in many countries. Across the European Union, however, member states' arrangements will share some similarities, as they have been designed to comply with the **Waste Framework Directive**. A good knowledge of a country's environmental legislation is required in order to effectively manage waste in that country.

8.3 Managing Waste

Disposal

Disposal must be to a permitted site, such as a landfill, incinerator or treatment works. It is important that you understand where waste is being taken for disposal and ensure the carrier is registered with a competent authority (if required by law in the country of transfer) to take the waste you produce. Disposal options are explained later.

Producer Responsibility

Producer responsibility is a policy tool that is an extension to the concept of 'polluter pays' that we considered earlier in the course. Its focus is to place responsibility for products when they get to the end of their life on the organisation that places the product on the market.

A former EU Environment Commissioner, Ritt Bjerregaard, described the aim of the concept as follows:

"Many manufacturers have for too long considered the problems of waste management as if it was somebody else's problem. It is important to clearly underline that it is also their problem. We cannot come to terms with the ever-growing amounts of waste in a rational way, unless concerns for waste minimisation and recovery are built into the product from the start."

Packaging Waste

Directive 94/62/EC on Packaging and Packaging Waste covers packaging that is placed on the market in the EU. Member states must take measures to stop the production of packaging waste, including programmes to encourage the re-use of packaging. The Directive also sets out a number of essential requirements for packaging, such as requirements specific to the:

- Manufacturing and composition of packaging.
- Re-usable nature of packaging.
- Recoverable nature of packaging.

Packaging must also comply with a 100ppm limit by weight for cadmium, lead, hexavalent chromium and mercury.

Similar requirements are covered in the UK by the **Producer Responsibility Obligations (Packaging Waste) Regulations 2007** and the **Packaging (Essential Requirements) Regulations 2015**.

Electrical and Electronic Waste

Waste Electrical and Electronic Equipment (WEEE)

Directive 2012/19/EU on Waste Electrical and Electronic Equipment (WEEE) aims for the prevention of WEEE. If this cannot be achieved, re-use, recycling and other forms of recovery of such wastes should be undertaken so as to reduce the disposal of such waste (re-use of WEEE as whole appliances is favoured over treatment, recycling and recovery).

The Directive also seeks to improve the environmental performance of all operators involved in the life cycle of EEE. Ultimately, the aim is to minimise the quantity of such items ending up in landfill. The target is for member states to collect 20kg per person per year, on average.

Electrical waste

All EEE placed on the market falls into scope unless specifically exempt or excluded. Categories of EEE identified in the Directive include:

1. Large household appliances.
2. Small household appliances.
3. IT and telecommunications equipment.
4. Consumer equipment and photovoltaic panels.

5. Lighting equipment.
6. Electrical and electronic tools.
7. Toys, leisure and sports equipment.
8. Medical devices.
9. Monitoring and control equipment.
10. Automatic dispensers.

Key requirements of the Directive include the following:

- Member states developing and maintaining a register of EEE producers.
- Householders must be able to take WEEE to a collection facility at no cost.
- Developing targets for the WEEE collected separately from households.
- Distributors and retailers are responsible for making arrangements to take back WEEE for free, in a way that is convenient for the customer.
- Introduction of recovery and recycling targets for WEEE for various categories.
- Producers to mark EEE products with the 'crossed-out wheeled-bin' symbol.

Similar requirements are covered in the UK by the **Waste Electrical and Electronic Equipment Regulations 2013**.

Restriction on the Use of Hazardous Substances

The aim of **Directive 2011/65/EU on the Restriction of the Use of Certain Hazardous Substances in Electrical and Electronic Equipment** (known as RoHS) is to contribute to the protection of human health and the environmentally sound recovery and disposal of WEEE.

The **RoHS Directive** covers all types of electronic equipment or goods dependent on electric current or electromagnetic fields for at least one intended function. However, a list of exclusions to the requirements of the regulations is provided in the Directive. Examples of such exclusions include:

- Equipment designed to be sent into space.
- Active medical implants.
- Photovoltaic panels (for public, commercial, industrial or residential use).designed to be sent into space.
- Large-scale stationary industrial tools.
- Transport for people or goods.

New EEE put on the market must not contain more than the permissible maximum concentration values of hazardous substances. These are:

- 0.1% by weight in homogeneous materials for lead.
- 0.1% by weight in homogeneous materials for hexavalent chromium.
- 0.1% by weight in homogeneous materials for mercury.
- 0.1% by weight in homogeneous materials for polybrominated biphenyls.
- 0.1% by weight in homogeneous materials for polybrominated diphenyl ethers.
- 0.01% by weight in homogeneous materials for cadmium.
- 0.1% by weight Bis (2-ethylhexyl) phthalate (DEHP).
- 0.1% by weight Butyl benzyl phthalate (BBP).
- 0.1% by weight Dibutyl phthalate (DBP).
- 0.1% by weight Diisobutyl phthalate (DIBP).

8.3 Managing Waste

> **DEFINITION**
> **PHTHALATE**
> Phthalates are chemicals that are added to plastics to make them flexible and supple.

Similar requirements are covered in the UK by the **Restriction of the Use of Certain Hazardous Substances in Electrical and Electronic Equipment Regulations 2012**.

Waste Batteries

Directive 2006/66/EC on Batteries and Accumulators and Waste Batteries and Accumulators (accumulators are rechargeable batteries) has the following requirements:

- The use of cadmium and mercury is prohibited above certain limits in batteries (this varies for different battery types; some battery applications have exemptions).
- Specific labelling is required to facilitate recycling (the 'crossed-out wheeled bin' symbol; 'Pb', 'Cd', 'Hg' if it contains lead, cadmium or mercury, respectively).
- Appliances that use batteries are designed so that the batteries can easily be removed.
- Battery producers have to register with the regulator, join and finance a battery compliance scheme (which will carry out waste battery collection, treatment and recycling obligations).
- Portable battery sellers have to take back waste (i.e. spent) portable batteries free of charge but may pass these on to a battery compliance scheme.
- Waste industrial and automotive batteries must not be disposed of by landfill or incineration.

Similar requirements are covered in the UK by the **Waste Batteries and Accumulators Regulations 2009** and the **Waste Batteries (Scotland) Regulations 2009**.

Waste from Construction Projects

Waste from construction projects can cause numerous environmental impacts if it escapes. Examples include pollution to air, land or water, fire hazards, threat to human health (e.g. asbestos wastes) and the impacts of landfilling or other disposal or treatment techniques. A site waste management plan is often produced for large construction projects. Such a plan will identify:

- Who is responsible for resource management.
- The types of waste that will be generated.
- How waste will be managed at each stage (with regard to the waste hierarchy).
- Which contractors will be used to ensure legal and responsible recycling and/or disposal.
- How the quantities of waste will be measured.

> **STUDY QUESTION**
> 4. Identify three ways in which waste could escape from control.
>
> (Suggested Answer is at the end.)

Outlets Available for Waste

IN THIS SECTION...

- A circular economy considers waste as a resource.
- Landfill sites and incinerators often require an environmental permit.
- Landfill sites can create a number of environmental impacts.
- Incineration can cause air pollution.
- In some countries, a tax is placed on all wastes going to landfill.

Circular Economy

A circular economy can be defined as an economy where resources are retained in use for as long as possible. During use the maximum value of the product is extracted and when the product comes to the end of its life, then as much material as possible is recovered. Such recovered materials can be used to make new products.

A circular economy differs from a linear economy, where the raw materials for a product are extracted, the product is manufactured and then used. When the product comes to the end of its use phase it is disposed of either by landfill or incineration without energy recovery.

The difference between the two types of economy can be summarised in the following two diagrams:

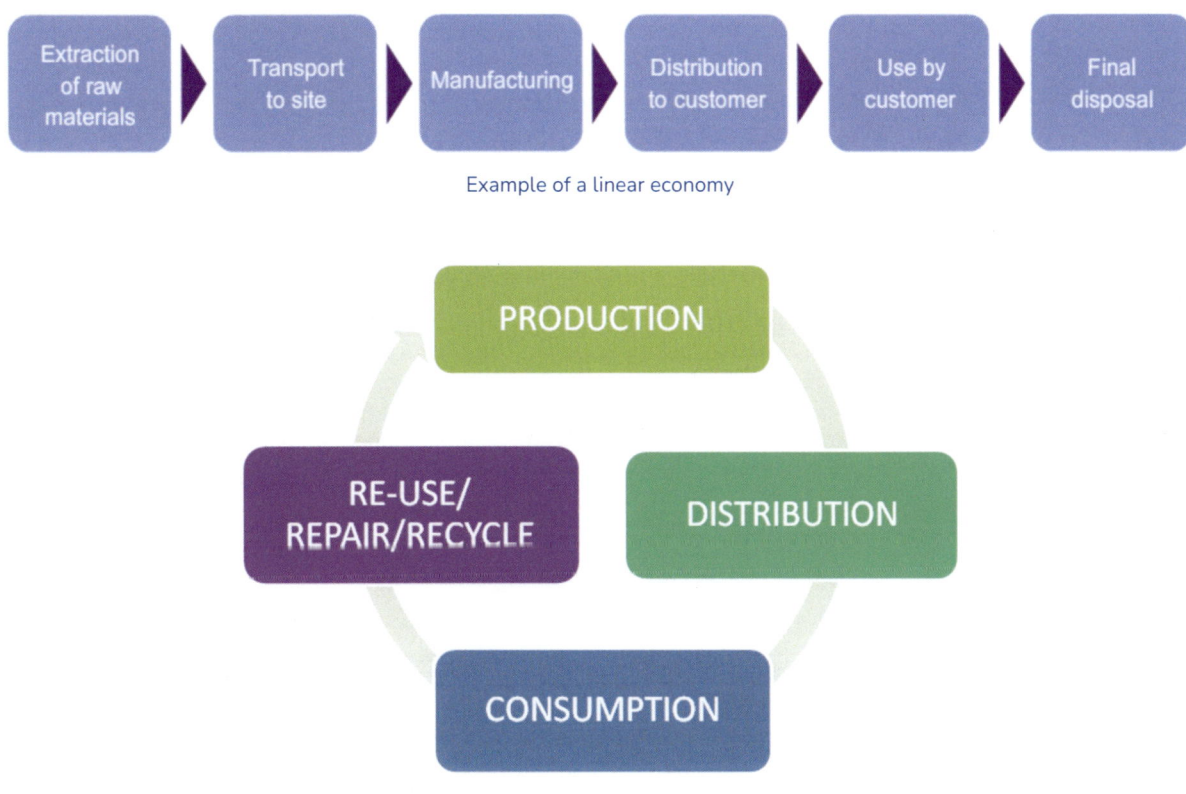

Example of a linear economy

Example of a circular economy

8.4 Outlets Available for Waste

The benefits of operating a circular economy in comparison to a linear economy are:

- Resources are kept in use for as long as is possible.
- The maximum value of each resource is extracted while in use.
- Products are recovered or regenerated at the end of their life.
- Reduction in pollution and waste.
- Delivers a more competitive economy.
- Reduces the environmental impacts associated with resource

Landfill and Incineration as Ultimate Disposal Routes

(See also Element 9 on energy recovery.)

Landfill

Landfill site management must adequately control and minimise any emissions, nuisances and litter, and this has implications during filling and afterwards. **Directive 1999/31/EC on the Landfill of Waste** (known as the **Landfill Directive**) requires that permitted landfill sites in the EU and UK are:

- Geologically suitable.
- Impervious, e.g. clay or sandstone, but if not a membrane can be laid to make it so.

The design should:

- Minimise ingress of groundwater, which could arise due to changes in the local water table, flood conditions or the existence of springs or streams.
- Be finished with an impervious clay dome to shed rainwater.
- Allow for a system of porous pipes to be laid to collect and encourage drainage of landfill leachate.
- Allow any leachate which accumulates to be pumped to a foul sewer following analysis to determine its constituents. If hazardous contaminants, typically toxic metals, are present, pre-treatment will be required. Leachate management can be a problem with older sites which were not carefully designed; it may end up contaminating groundwater or invading local streams, especially in times of flood.

Landfill

> **DEFINITION**
>
> **LANDFILL LEACHATE**
>
> Is liquid that drains or moves through a landfill. The liquid is either already in the waste or is caused by rainfall. It is often highly contaminated and can pollute nearby waterways and groundwater.

Nuisances must be adequately controlled, especially if the site is near a populated area:

- **Noise nuisance:** heavy vehicle movements, both site traffic bringing waste for disposal and site plant. Permit conditions may specify working hours.
- **Odours**: minimised by:
  - the cell method of filling;
  - ensuring the surface is covered with inert fill at the end of each working day; and
  - operating to minimise exposed areas.

  Chemical sprays to mask smells may be used in unusual wind conditions.

8.4 Outlets Available for Waste

- **Dust and litter:** minimised by damping down and good site practice, e.g. cell filling by tipping down a gradual slope and using specially designed plant to bury litter and increase the fill density, maximising the site capacity and minimising later settlement. This also reduces the possibility of fire (producing smoke, smells and contaminated leachate) starting in the waste.
- **Vermin:** gulls, rats, mice and foxes. Good site practice is required, and possibly an eradication programme for rats.

Site security with a chain-link fence of suitable height will keep unauthorised persons out and help to catch wind-blown litter.

Landfill Gas (LFG) is a combustible mix of methane and carbon dioxide which can:

- Present explosion and toxic hazards to premises near the site by permeating through soil.
- Injure site-screening trees.

Municipal Solid Waste (MSW) contains sufficient putrescible material to give a good supply of landfill gas - production may continue for many years. Such gas may be odourless and a detailed risk assessment should be undertaken, along with a programme of monitoring. Landfill gas is normally collected in pipes laid within the waste and is either flared off (burnt) or collected and used as fuel.

> **DEFINITION**
> **PUTRESCIBLE**
> Capable of rotting/decomposing.

Landfill acceptance criteria require waste producers and treatment businesses to decide what level of treatment to pursue, according to the type of waste and the category of landfill it is to go to. Landfill sites are categorised according to the type of waste they are permitted to take, as follows:

- Inert waste sites.
- Non-hazardous landfills.
- Hazardous waste landfills.

8.4 Outlets Available for Waste

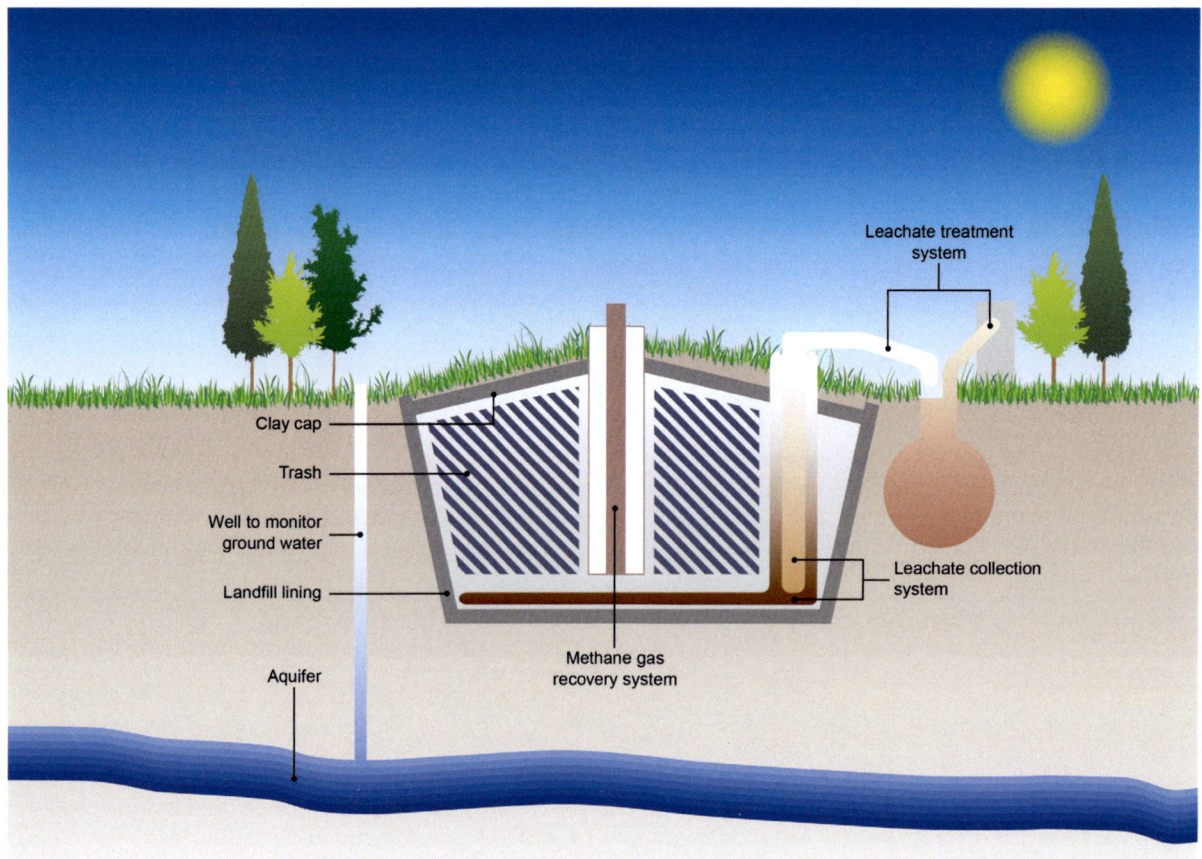

Section through a typical landfill site

Incineration

Incineration with energy recovery requires:

- A capital-intensive plant.
- Detailed planning.
- An EIA.

MSW has about 40% of the calorific value of coal and is therefore a considerable source of energy. The BPEO (if you are not sure what this means, look back at Element 1) for an incinerator project is usually **Combined Heat and Power (CHP)**. Waste reduction is 60-90% and results in clinker (usable for road-building), fly ash (landfilled) and metal (recycled).

In the UK, the Edmonton incinerator was designed to burn 400,000 tonnes per annum of MSW from five London boroughs. It takes 1,330 tonnes of refuse each day, produces superheated steam and drives four turbines supplying electricity to the National Grid, earning £4m per annum.

The temperature of combustion in the incinerator is controlled between 925 and 1,040°C; complete combustion is achieved by agitation of the burning waste turning and falling from roller to roller in the incinerator grate. From the final roller, burnt-out clinker and ash pass through a quench bath onto conveyors and are carried to the residuals handling area. Exhaust gas from the economiser passes through 50kV electrostatic precipitators before discharge to the atmosphere from the twin flues of the main chimney. The precipitators reduce the dust content of the gases to less than 60mgm^{-3}.

TOPIC FOCUS

Advantages and disadvantages of landfill and incineration

	Advantages	Disadvantages

Landfill	• Comparatively cheap disposal.
	• Can be used to restore areas used for quarries, etc. to local amenity use.
	• Able to take large volumes of waste.
	• Modern landfill is of a low risk for humans and the environment.
	• Can be used to generate electricity.
		• Cheap, so does not encourage producers and consumers to migrate up the waste hierarchy.
		• Waste does not break down quickly, so the problem of management remains for a considerable time after a site has closed.
		• Poorly managed sites can lead to:
		   – Surface and groundwater pollution from leachate.
		   – Air pollution from unmanaged LFG or fires.

Incineration	• Reduction in volume. • Allows heat to be recovered. • An energy source.	• High initial cost. • Volume of traffic. • Monitoring of air pollution. • Disincentive to recycling schemes (incinerators may need constant feeding).

Other Treatment or Disposal Routes

As well as landfill and incineration, there are other routes to disposal, including:

- **Domestic waste sites** - allow local residents to dispose of unwanted items that cannot be disposed of through the usual household collection system. Many now operate separate collections for many types of waste, such as used oil, cardboard, clothes, paper, metal, glass, etc. that feed into the local authority recycling process.
- **Waste transfer stations** - because many landfill sites and incinerators are long distances from where the waste is collected, most waste will pass through a waste transfer station. Local waste collection vehicles can carry about 10 tonnes of waste. This is 'bulked up' onto a larger vehicle with a payload of approximately 20 tonnes, before being transported to the point of final disposal, such as a landfill site. This reduces the volume of vehicles on the road and these larger vehicles are usually better adapted for driving on the landfill area. Many transfer stations are now operating as Material Recovery Facilities (MRFs) as well. This involves separating from the waste stream any materials that can be recycled and diverting them from landfill.
- **Waste treatment facilities involving recovery operations** - recovery operations may include:
  - Simple sorting of waste (as described above for MRFs) to divert recyclable and re-usable materials away from landfill/incineration.
  - Gas/heat recovery and electricity generation from landfill and incineration activities (as discussed earlier).

8.4 Outlets Available for Waste

Global Waste Trade

Waste disposal in some countries can sometimes be poor, presenting a major risk to people's health and the environment. Increasing amounts of imported waste and a change in composition are a key challenge to local governments.

Often, the cost of safe disposal is beyond their financial means; there is also often a lack of capacity and political will to tackle the issue. Therefore, the main way of dealing with waste is uncontrolled dumping. With the growth of industry and trade, the amount of hazardous industrial waste will rise, posing an even greater risk. Countries that export waste have a significant amount of responsibility to ensure that waste is disposed of in a manner that does not impact the environment in other countries. Indeed, the Basel Convention on the Control of Transboundary Movements of Hazardous Wastes and their Disposal regulates the import and export of hazardous waste and certain plastic wastes.

Export Costs and the Impact of Export, Landfill and Aggregate Taxes

Taxes can be applied to waste, with the aim of reducing the volume of waste going to landfill and making alternative disposal options more financially viable. Export costs for waste shipped internationally may also apply.

The details of such taxes will vary between countries; as an example, specific taxes applicable in the UK are outlined below.

Landfill Tax

The Landfill Tax was introduced in the UK in 1996 to impose a tax on all waste disposed of at permitted landfill sites after 1 October 1996. Landfill Tax is chargeable on waste from both permitted/licensed and non-permitted/licensed sites such as those where illegal disposal of waste has occurred.

There are two rates of LFT:

- One for **inert wastes**, such as:
    - Naturally occurring rocks and soils, e.g. clay, sand and gravel.
    - Ceramic or cemented materials, such as glass, ceramics, bricks, pottery and china.
    - Ash from wood or coal combustion.
    - Gypsum and calcium-sulphate-based plaster, providing it is disposed of in a separate containment cell or in an inert landfill site.
    - Furnace slags.
- One for **active wastes**, such as:
    - Food and kitchen wastes.
    - Garden waste.
    - Packaging waste.
    - Agricultural waste.

Some wastes are usually exempt from LFT, for example:

- Naturally occurring minerals from mining and quarrying.
- Burial of domestic pets at pet cemeteries.
- Materials used in the engineering of the landfill site, such as:
    - Clay.
    - High Density Polyethylene (HDPE) used to line the base of the landfill.
    - Tyres used to create a drainage layer at the bottom of the cell.

The current rates for LFT are £3.15 per tonne for inert wastes and £98.60 per tonne for active waste.

The Environmental Bodies Credit Scheme was set up to encourage landfill site operators to fund projects that benefit the environment in England and Northern Ireland. Under the scheme, landfill site operators can claim a tax credit for contributions made to approved environmental bodies. These bodies are enrolled by Entrust, the regulatory body specifically set up to oversee the scheme. Landfill site operators may claim a tax credit worth 90% of any contribution made to an enrolled environmental body for spending on an approved object, subject to a maximum credit of 5.3% of the landfill tax paid during the year. Similar systems are in place for landfill tax in Scotland and Wales.

Aggregates Levy

Also in the UK, the **Finance Act 2001** enabled the Aggregates Levy to be introduced, which came into effect on 1 April 2002. The purpose of the tax is to:

- Address, by taxation, the environmental costs associated with quarrying operations (noise, dust, visual intrusion, loss of amenity and damage to biodiversity).
- Reduce demand for aggregates and encourage the use of alternative materials where possible.

The current rate for the Aggregates Levy is £2.00 per tonne.

STUDY QUESTIONS

5. List four nuisances that must be adequately controlled on a landfill site, particularly if it is near a populated area.
6. Give three advantages and three disadvantages of:
   (a) Landfill.
   (b) Incineration.

(Suggested Answers are at the end.)

8.5 Risks Associated with Contaminated Land

Risks Associated with Contaminated Land

IN THIS SECTION...

- Land contaminated by industrial processes can have a number of harmful effects to humans, plants, animals and water.
- There are significant liabilities associated with contaminated land.

The Potential Effects of Contaminated Land on the Environment

Land contaminated by industrial processes can have direct and indirect effects on human health. Contaminated land also brings with it significant liabilities for the owners of the land, even if they are not the producer of the contamination.

Many industries have a significant potential to contaminate the land on which they operate. Some of the most common ones are:

- Mining and extractive industries.
- Iron and steel works.
- Oil refining and storage.
- Sewage treatment works.
- Chemical and pharmaceutical works.

Contaminated land

> **TOPIC FOCUS**
>
> **Effects of Contaminated Land**
>
> The **hazards** associated with land contamination include:
>
> - Direct risk to human health by ingestion, particularly of toxic metals that accumulate in the body, e.g. lead and cadmium.
> - Toxic substances entering the food chain through plant uptake.
> - Water contamination:
>   - Directly by migration through plastic water pipes (especially of phenols and cresols) into household water supplies.
>   - Indirectly by leaching into groundwater.
> - Prevention or inhibition of plant growth:
>   - Phytotoxic (plant poisoning) contaminants may affect plant growth at low concentrations (e.g. toxic metals).
>   - Gases may displace air from soil and impede growth.
> - Odours, fumes and effluvia, particularly when gases (landfill gases) percolate the site.
> - Fire and explosion:
>   - Gas explosions from methane and petroleum vapours.
>   - Underground fires (responsible for odour, fumes, etc. and ground subsidence).
>
> (Continued)

Risks Associated with Contaminated Land | 8.5

> **TOPIC FOCUS**
> - Direct contact with, or inhalation of, certain contaminants - solvents, corrosives, carcinogens, asbestos, etc. Inhalation of dust, vapours (phenols and other aromatic compounds) and toxic gases may add to exposure. Skin contact with coal-tar residues can cause dermatitis, tar warts and skin cancer.
> - Building damage, e.g. high levels of sulphate can occur downwind of brickworks, along with fluorine as dry deposition on land over many years.

When considering health effects to humans, there are a number of different pathways a contaminant can take, depending on the type of contaminant.

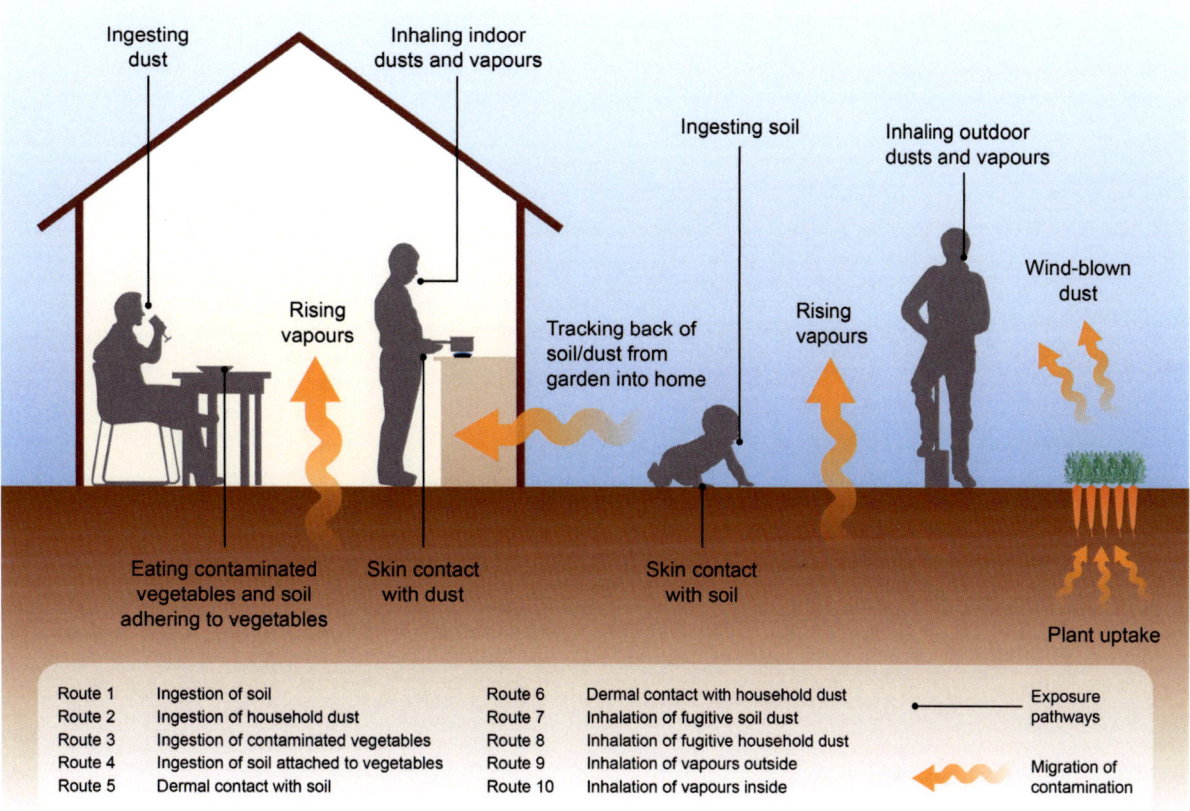

People may be exposed to contaminants by a number of pathways

Contaminated Land Liabilities

Contaminated land, and any knock-on effects of the contamination, such as pollution of groundwater resources, can be difficult and expensive to remediate. Much land contamination has been caused historically, where the organisation responsible for causing the contamination no longer exists. Determining liabilities for remediation may therefore be complex.

The legal basis for liability for contaminated land in the UK is contained in Part 2A of the **Environmental Protection Act 1990**. Local authorities are required to make a strategic assessment of likely contaminated sites and to undertake more detailed inspections of the presence of contamination based on this. The authority must then undertake a risk assessment to understand the risks posed and must designate a site as being contaminated if:

- Significant harm is being caused to a human, or relevant non-human, receptor.
- There is a significant possibility of significant harm being caused to a human, or relevant non-human, receptor.

8.5 Risks Associated with Contaminated Land

- Significant pollution of controlled waters is being caused.
- There is a possibility of significant pollution of controlled waters being caused.

Once land has been determined as contaminated land, the enforcing authority must consider how it should be remediated and, where appropriate, it must issue a remediation notice to require such remediation. The enforcing authority for the purposes of remediation may be the local authority which determined the land, or the Environment Agency.

Liability for complying with the remediation notice is determined taking into account a number of factors:

- The possibility of identifying the person (individual or corporation) responsible for causing the contamination.
- The history of contamination of the site - contamination may have been caused over different time periods, by a number of different persons.
- The current owner or owners of the land.

Understandably, many contaminated sites are complex and liability for remediation may be assigned to a 'liability group' of persons who are deemed to share some measure of individual responsibility.

MORE...

The UK's DEFRA has produced statutory guidance on contaminated land which may be found at:

www.gov.uk/government/uploads/system/uploads/attachment_data/file/223705/pb13735cont-land-guidance.pdf

STUDY QUESTIONS

7. In relation to land contamination, give an example each of how water can be directly and indirectly contaminated.
8. List the hazards associated with land contamination.

(Suggested Answers are at the end.)

Summary

This element has dealt with the control of waste and contaminated land.

In particular, this element has:

- Outlined categories of waste, including:
  - Inert waste - waste which is 'stable', e.g. uncontaminated bricks, glass, concrete, etc.
  - Hazardous waste - any waste that has a hazardous property (e.g. corrosive, carcinogenic or ecotoxic).
  - Non-hazardous waste - controlled waste that does not exhibit properties required to define it as hazardous.
  - Clinical waste - surgical swabs, dressings, animal tissue, etc.
  - Radioactive waste - waste that contains or is contaminated with radionuclides at concentrations or activities greater than clearance levels, as established by the regulatory body.
  - Controlled waste - waste controlled by relevant legislation in the UK.
- Explained the waste hierarchy: prevent its production, prepare products/components for re-use, recycle/compost, recover energy after some form of treatment, and responsible disposal.
- Shown how waste needs to be controlled all the way through the waste chain, from its source to site of final disposal.
- Outlined how waste should be appropriately separated, stored, identified and labelled at all times.
- Explained that waste should be transferred only to a waste carrier who is registered with a competent authority.
- Outlined how, in the EU, **Directive 94/62/EC on Packaging and Packaging Waste** and **Directive 2012/19/EU on Waste Electrical and Electronic Equipment (WEEE)** operate to prevent waste formation.
- Shown that construction sites generate significant quantities of waste.
- Described the outlets available for waste, including:
  - Landfill and incineration, both of which have advantages and disadvantages.
  - Other treatment and disposal methods, such as waste transfer stations and treatment facilities involving recovery operations.
- Outlined how taxes can be applied to waste, with the aim of reducing the volume of waste going to landfill and making alternative options for disposal more financially viable (using UK taxes as an example).
- Explained that many industries contaminate the land on which they operate (e.g. mining and extractive industries, iron and steel works, oil refining and storage).

Exam Skills

Question

Scenario
You have been tasked with developing an action plan for eliminating waste to landfill (zero waste to landfill) from the organisation for which you work. The facilities team are sceptical about the development of the plan as they are under the impression that landfilling waste is an easy and cheap option.

Task
What are the reasons as to why your organisation should comply with the zero waste to landfill commitment? **(10 marks)**

Approaching the Question

Think about the steps you would take to answer the question:

- Read the scenario carefully. With this question you need to develop some reasons as to why no longer disposing of the company's waste to landfill is a good idea.
- Now look at the task – prepare notes on:
  - The landfilling of waste.
  - The advantages of using alternative techniques for waste.
- Consider the marks available. In this case, there are 10 marks available so you should provide 10 reasons as to why the organisation should comply with the zero waste to landfill commitment.
- Read the scenario and task again to make sure you understand them and have a clear understanding of the reasons for the zero waste to landfill commitment.
- Jot down an outline plan - this might include:
  - Climate change, health and safety, water/groundwater, resources, land take, development restrictions, community issues, regulation, transport, lack of access to landfill, EMS, pre- treatment.

Now have a go at the question yourself.

Exam Skills

Example of How the Question Could be Answered

There are many reasons why producers of waste are looking at reducing the use of landfill as a disposal route, including:

- Climate-change impacts such as methane emissions from the breakdown of waste and carbon dioxide emitted from transportation and other activities.
- Health and safety implications of explosive and flammable landfill gas.
- Water and groundwater pollution from the production of leachate.
- Placing resources underground that could be re-used, recovered or recycled.
- Land take for landfill sites, including restriction on the activities the land can be used for following closure of the site.
- Restriction on development of the land due to health, safety and environmental concerns for buildings that are in the vicinity of landfill sites.
- Concerns of local communities and impacts on organisations of lobbying by environmental groups.
- Increased cost due to landfill tax and stricter regulatory control.
- Greater transport distances due to pattern of fewer larger sites.
- New sites development being restricted due to planning permission.
- Management system requirements such as ISO 14001, meaning that other options for wastes need to be considered, e.g. re-use, recovery, recycling, and other disposal methods.
- Waste for landfill must be subject to pre-treatment.

Element 9

Sources and Use of Energy and Energy Efficiency

Learning Outcomes

- Understand the benefits and limitations of a range of energy sources, and suggest suitable measures to increase energy efficiency.

Learning Objectives

Once you've read this element, you'll be able to:

1. Discuss the benefits and limitations of a range of renewable and non-renewable energy sources.

2. Explain how energy efficiency can be increased.

Contents

Use of Fossil Fuels — 9-3
Examples of Fossil Fuels — 9-3
Benefits and Limitations of their Use as an Energy Source — 9-6
Carbon Offsetting — 9-7

Renewable Sources of Energy — 9-9
Fossil Fuel Alternatives — 9-9
Benefits and Limitations of the Use of Alternative Energy Sources — 9-14
Energy Supply in Remote Locations and Developing Countries — 9-16
On-Site Energy Generation and Storage — 9-16
Benefits and Limitations of Using Emerging Technologies for Energy — 9-16

Energy Efficiency — 9-17
Benefits of Energy Efficiency — 9-17
Energy Monitoring — 9-17
Control Measures Available to Increase Energy Efficiency — 9-18
Fuel Choice for Transport and the Optimisation of Vehicle Use — 9-21

Summary — 9-24

Exam Skills — 9-25

Use of Fossil Fuels

IN THIS SECTION...

- Fossil fuels are formed from the organic remains of marine micro-organisms (oil and gas) and land-based vegetation (coal).
- Advantages of fossil-fuel use:
    - Easy combustion and transportation.
    - Fuels are relatively inexpensive.
    - Generation of electricity is efficient and inexpensive.
    - Power stations can be built anywhere.
- Disadvantages of fossil-fuel use:
    - Environmental effects, including acid rain, climate change and damage from extraction.
    - Fuels are non-renewable.
    - Prices are variable depending on markets.
    - Emissions contribute to poor air quality.

There are benefits but also disadvantages to carbon offsetting.

Examples of Fossil Fuels

Coal, oil and gas are called fossil fuels for the simple reason that they are formed from the fossilised organic remains of plants:

- Marine micro-organisms in the case of oil and gas.
- Land-based vegetation in the case of coal.

The processes that gave rise to the current stores of oil, gas and coal began in the Carboniferous period of the Paleozoic Era some 360 to 286 million years ago.

Oil is called a fossil fuel because it is formed from the fossilised organic remains of plants

Formation of Oil and Gas

Oil and gas are derived from the accumulation of very large quantities of organic material on the floor of ancient oceans. The organic material mostly consisted of the remains of microscopic plants that lived in the upper layers of the oceans and captured carbon from atmospheric carbon dioxide (CO_2) dissolved in seawater through the process of photosynthesis. In some parts of the oceans, where the sea-floor was stagnant and deprived of oxygen, the remains of these dead organisms were not readily broken down, allowing layers to accumulate. A high rate of sedimentation then ensured that the accumulations of dead organisms became rapidly buried.

As the sedimentation processes continued, the plant remains were subject to increasingly high pressures and temperatures, converting the carbon contained in the dead tissues into oil and gas. Over time, these hydrocarbons migrated through porous rocks, such as sandstones, and accumulated in areas that were overlain by impervious strata. These pockets of hydrocarbons formed the oil and gas fields that we exploit today.

9.1 Use of Fossil Fuels

Formation of Coal

Coal is formed through a very similar land-based process. Vegetation in low-lying, swampy conditions is the most likely to form deposits of coal. As with the production of oil and gas, a stagnant environment excludes oxygen and so reduces the breakdown of dead vegetation, allowing it to accumulate and to form peat. As this process continues, successive layers of peat are buried and eventually covered by sediments. This process of continual burial with the addition of heat over long periods of time, breaks down and alters the hydrocarbons in the peat. The peat then goes through a number of stages, becoming richer in carbon at each stage, as other elements are dispersed from the original material. These stages are:

- Peat.
- Lignite.
- Sub-bituminous coal.
- Bituminous coal.
- Anthracite coal.
- Graphite (pure carbon).

Common Fossil Fuels

The fossil fuels that are in common use today are:

Fuel	Description
Coal	Solid, black carbon-based mineral.
	The largest source of fuel for electricity generation worldwide.
Anthracite	Very high-quality coal, with a carbon content of >90%.
Lignite	Low quality 'brown coal', with a carbon content of 60-70%.
Coke	Solid residue made by heating coal in the absence of air. Widely used in steel furnaces.
Petrol/Gasoline	Mixture of hydrocarbon compounds distilled from petroleum (crude oil) and containing various additives. The most common fuel for passenger road vehicles and aircraft.
Diesel/DERV	Mixture of hydrocarbon compounds distilled from petroleum but more dense than petrol.
	The most common fuel for freight road vehicles.
	Sometimes used to fuel boilers for space heating.
	(Note that biodiesel is not a fossil fuel.)
Fuel oil/Heating oil/Paraffin	Mixtures of hydrocarbon compounds distilled from petroleum.
	Available in various grades.
	Used to fuel domestic and industrial boilers and ships.
Natural gas	Naturally occurring hydrocarbon gas mostly consisting of methane.
	Widely used for domestic and industrial heating and electricity generation.
Liquefied natural gas	Natural gas condensed to a liquid for bulk transport by refrigerating to -162°C.
	Converted back to gas before use as above.
Liquefied petroleum gas	Propane or butane (or a mixture).
	Used as a fuel in heating appliances, cookers and sometimes road vehicles.

Use of Fossil Fuels | 9.1

In certain countries, such as the Republic of Ireland, peat is also extracted on an industrial scale for use in electricity generation.

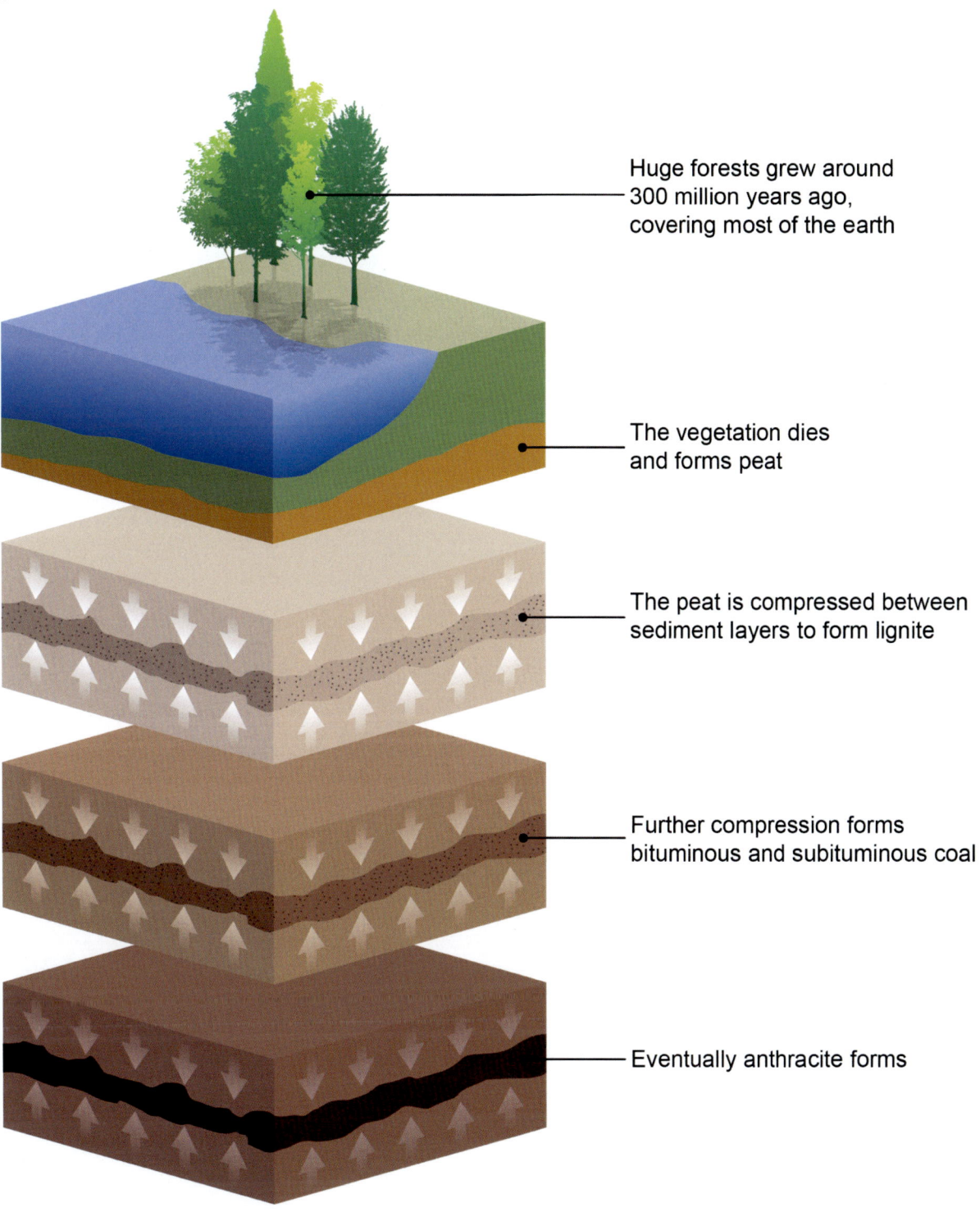

Fossil-fuel formation on land

9.1 Use of Fossil Fuels

Where Supplies Come From

Very specific geological conditions are required for the formation of oil, gas and coal, so it is not surprising that significant deposits of fossil fuels are found in specific areas around the world, such as the Middle East, Russia and the United States. However, it is not just a question of where the deposits are that influences where the supplies come from. Other aspects are also significant, such as:

- Ease of access to the oil, gas or coal.
- How it can be transported from source to processing factory and finally to the consumer.

Clearly, oil and gas have an advantage over coal in that they can be transported by pipeline.

Benefits and Limitations of their Use as an Energy Source

Benefits

Since fossil fuels were first discovered, they have provided an ever increasingly important supply of energy. Our modern lifestyle would be impossible to maintain without the benefits that fossil fuels have provided. Coal was probably the first fossil fuel to be used regularly as a replacement for wood in cooking fires. In many areas of the world, low-grade coal could be found near to the surface and was therefore easy to access. The high-grade coals only became generally accessible with the invention of the steam-driven pump that could de-water deep mines where the better coal was to be found. Coal is still a major provider of energy for both domestic and industrial use. It is used in power stations to generate electricity and in manufacturing processes, such as cement manufacture.

Oil has a wide range of uses, such as energy production, and is a raw material for many everyday goods, such as plastics. Oil enables:

- Goods to be moved around the globe economically, so providing an ever-increasing range of consumer products for us to purchase.
- Access to cheap fuel, allowing many people to own and drive a car and have good quality food all year round in their local supermarket.

Compared to an alternative power source, such as nuclear, oil and coal are low risk and simple technologies. It is easy to build a coal- or oil-fired power station and relatively easy and cheap to transport the required fuel from point of production to point of use.

Limitations

Although it is clear that there are many advantages to using fossil fuels, they are not without their limitations:

- The burning of fossil fuels is now believed to be a major contributor to climate change through the emission of large volumes of CO_2 to the atmosphere.
- Fossil fuels are known to contribute significantly in some areas to the production of acid rain that leads to damage to flora and fauna over large areas of land.
- The extraction of the raw materials often leads to significant damage to the environment.
- Oil spills from pipelines and tankers often impact highly sensitive areas for wildlife.

The table below summarises the advantages and disadvantages of the use of fossil fuels.

Warning of oil spill on beach

Use of Fossil Fuels — 9.1

Advantages and disadvantages of fossil-fuel use

Advantages	Disadvantages
Straightforward combustion process.	Major contributor to climate change.
Relatively inexpensive.	Cause acid rain.
Easily transported.	Non-renewable resources that are not sustainable in the long term.
Large amounts of electricity can be generated in one place, quite cheaply.	Prices are susceptible to changes in global politics so may rise significantly at short notice.
Gas-fired power stations relatively efficient.	Extracting the raw materials can be dangerous and damaging to the environment.
Power stations can be built almost anywhere.	Emissions may contribute to poor air quality locally, thereby affecting people's health.

Not all fossil fuels are equal in their impact on the environment:

- Gas burns cleaner than oil and produces about twice the energy per kilo and proportionately less CO_2 when burnt.
- Different oil products also contribute differently to air pollution when burnt, e.g. petrol burns comparatively cleanly against the heavy fuel oil burnt in the shipping industry.
- Heavy oil and diesel produce a high volume of particulate matter that can have an adverse effect on both plant and animal (including human) life.

Carbon Offsetting

Carbon offsetting is an activity that results in compensation for emissions of greenhouse gases by enabling emission reduction elsewhere. It is often used to enable carbon neutrality - where the carbon reductions are equivalent to the total amount of carbon emitted. Carbon offsets are usually purchased in the form of credits from projects where carbon reductions are occurring, such as:

- Renewable energy projects.
- Energy efficiency improvements.
- Destruction of potent greenhouse gases such as HFCs.
- Carbon sequestration projects such as tree planting.

The benefits and limitations of carbon offsetting are:

- **Benefits**
  - Business benefits - demonstrate green credentials, improve public relations and marketing.
  - Enable the counterbalancing of carbon emissions when it is not possible to avoid creating emissions.
  - May support a programme which can have other environmental, economic and social benefits.
- **Disadvantages**
  - Offsetting often detracts from solutions to reduce carbon emissions within an organisation.
  - Allows emitters to continue to release greenhouse gases.
  - The carbon offset market is largely unregulated and there is no agreed global standard.
  - Lack of transparency for some schemes.
  - It is hard to quantify the benefits of carbon offsetting, for example what amount of carbon dioxide would a tree take from the atmosphere? Would this be enough to cancel out the emissions from a business flight?

9.1 Use of Fossil Fuels

- It is difficult to give a monetary value on environmental damage as money has different value to different people.

STUDY QUESTION

1. List three advantages and three disadvantages of the use of fossil fuels.

(Suggested Answer is at the end.)

Renewable Sources of Energy

IN THIS SECTION...

- Alternative sources of energy include: solar, wind, hydroelectric, wave, tidal power, geothermal, nuclear, combined heat and power, biodigesters, methane recovery and biomass.
- There are both benefits and limitations to the use of alternative energy sources.
- Problems can occur with energy generation and supply in developing countries and remote regions.
- There are benefits and limitations of using emerging technologies for energy.

Fossil Fuel Alternatives

Alternatives to fossil fuel are being increasingly developed as energy sources for a number of reasons:

- There is concern over the effects of fossil-fuel use on the environment.
- Costs of fossil fuels are increasing.
- Demand for fossil fuels is starting to outstrip supply.

Here, we will look at some of the main alternative sources, although there are many other potential sources and many new ones being developed.

Solar

Effectively, all energy on the planet originates with solar energy. Even fossil fuels originally gained the energy they stored through photosynthesis and respiration via energy from the Sun. Humans have used direct solar energy for thousands of years to dry clothes and grow food. However, it is only recently that we have been using it to generate electricity. The Sun produces far more energy than we require, so if we can develop efficient ways to capture that energy, many problems could be resolved.

Photovoltaic cells

There are three main ways in which we use solar power:

- **Solar cells** (photovoltaic or photoelectric), where photovoltaic panels convert light into electricity at an atomic level through the use of materials that exhibit a property known as a photoelectric effect. This effect causes them to absorb photons of light and release electrons. When these electrons are captured, an electric current is generated. They are a common way of generating power for both domestic and commercial properties in many parts of the world.
- **Solar water heating** uses energy from the Sun to directly heat water in glass panels, thereby reducing the amount of energy from fossil fuels required to provide hot water for use in the house.
- **Solar furnaces** are commercial installations that use a large number of mirrors to concentrate the energy of the Sun into a small space and to allow the production of very high temperatures. Some of these furnaces can produce temperatures up to 33,000°C.

A significant drawback of solar energy is that it is only effective during the hours of daylight, whereas peak energy consumption is often after nightfall, especially in the winter months. Large surface areas are also required for photovoltaic arrays and, as with wind power (see below), this has led to concerns that renewable energy infrastructure is damaging the landscape and amenity value of previously undeveloped countryside.

9.2 Renewable Sources of Energy

Wind

The use of wind as an energy supply is also not new:

- Wind was in use by the Babylonians and Chinese as long as 4,000 years ago to pump irrigation water.
- Windmills used for grinding corn were common in Europe in the Middle Ages.

Modern windmills are actually wind-powered turbines. They use the power of the wind to generate electricity. Wind farms, as they have become known, are now a common sight.

Wind-Farm Operation

- There needs to be an average wind speed of 25km/h for them to operate efficiently.
- At very high wind speeds, many wind turbines become less efficient or need to shut down to prevent damage occurring.

Wind-powered turbines

- Propellers of modern wind turbines are much larger than traditional windmills and are mounted much higher. This is to ensure they capture energy from the largest possible volume of air.
- The blades can have their pitch adjusted to increase the range of wind speeds in which they can operate and to maximise efficiency.
- The body of the turbine can be turned so the blades are always facing into the wind.

Hydroelectric

The most common type of hydroelectric energy generation uses the flow of water under gravity to turn a turbine attached to a generator. The generator converts the rotary mechanical motion into electricity. The water is usually held in a reservoir behind a dam. Hydroelectricity is extensively generated in Scotland, which has an abundance of the required natural resources.

Other methods of hydroelectricity generation use undershot waterwheels, utilising the flow of the water below a vertically-mounted wheel (like a water mill, only the mechanical energy is converted to electricity rather than being used directly).

Wave

Wave power generation is still rare as it is difficult to harness the energy of the waves. One of the most common ways of generating power from waves is through the use of an air chamber and turbine connected to a generator. The following figure shows how the waves make the water in the chamber move up and down. The movement of air drives the wind turbine and that, in turn, drives the generator.

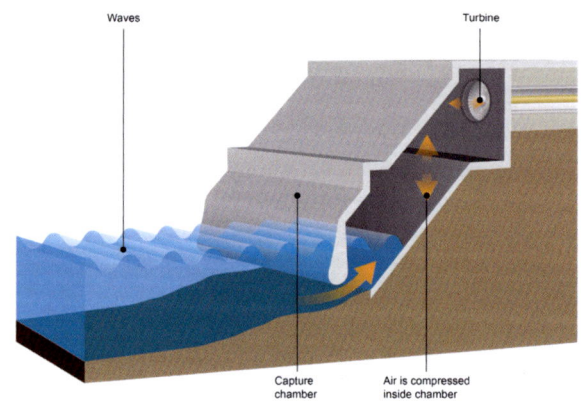

Example of a wave turbine generator

Tidal Power

Tidal power uses the energy of rising and receding tides to generate power. Although large amounts of power could be generated in this way, tidal power schemes are rare because the technological barriers are significant.

Tidal Power Operation

Tidal power works in the following way:

- The system is similar to a hydroelectric power station using the movement of water to drive a turbine.
- A dam or barrage needs to be built across a river estuary to focus the power of the tide.
- As the tide comes in, water moves from the seaward side to the estuary side of the barrage.
- The turbine inside the barrage is driven by the power of the water.

Geothermal

As we go down through the Earth, the temperature increases to a point where rocks become molten. In general, the temperature increases by 1°C every 36 metres in depth. The origin of the heat is continuous radioactive decay deep inside the Earth. It is possible to use this heat to generate electricity.

Geothermal Operation

Geothermal operation works in the following way:

- Holes are drilled down until they reach a depth where the temperature is hot enough to boil water.
- Pipes are then installed in these wells and water is pumped down the pipes.
- As the water turns to steam it drives a turbine that is connected to a generator and produces electricity.

Effective exploitation of geothermal energy is really only feasible in particular regions, where geological conditions are favourable; for example, volcanic areas such as Iceland.

Nuclear

Nuclear power uses the heat from radioactive processes to generate steam that is used to power turbine-driven electricity generators. Nuclear fission of radioactive isotopes of uranium is currently the most common source of energy in nuclear power stations, although some fuel uses other metals, such as plutonium. Nuclear power stations provided about 11% of the world's electricity in recent years.

> **DEFINITION**
>
> **NUCLEAR FISSION**
>
> This involves 'splitting the atom'. The energy used to hold the atom together is released in the form of heat.

An important advantage of nuclear power is that relatively small amounts of fuel produce a lot of power. Also, the output from nuclear power stations is very reliable and not weather-dependent, as is the case with other fossil-fuel alternatives, such as wind and solar.

One of the most compelling arguments in support of nuclear power is that it has a very low carbon footprint - each kilowatt-hour (kWh) of nuclear electricity has about the same level of carbon emissions as a kWh of electricity produced by wind power. Many observers therefore see nuclear as an important route to combating global warming.

The main drawback of nuclear power stations is that they generate significant volumes of potentially dangerous radioactive waste. Power station operations result in:

- Large volumes of waste with a low level of radioactivity, such as contaminated protective clothing and equipment.

9.2 Renewable Sources of Energy

- Significant volumes of waste with a high level of radioactivity from the re-processing of spent fuel. These typically contain radioactive isotopes of metals, such as technetium, iodine, neptunium and plutonium. These materials are highly dangerous to living organisms and have half-lives of tens of thousands, or even millions of years, which means that they require long-term secure storage.

Accidents at nuclear power stations also have the potential to release radioactivity into the surrounding environment. For example, mechanical failures can result in losses of cooling gases or liquids that are contaminated with radioactivity. More serious failures in reactor cooling systems can result in catastrophic fires that release very large amounts of high-level radiation that can be transported hundreds of miles in the atmosphere.

Although accidents at nuclear power stations are rare, a number of high-profile incidents have occurred, for example:

- Chernobyl, Ukraine, 1986: a catastrophic reactor fire sent a plume of highly radioactive fallout into the atmosphere that contaminated large areas of Belarus, Ukraine and Russia and led to the evacuation of around 350,000 people.
- Fukushima, Japan, 2011: the combined effects of an earthquake and a tsunami resulted in failure of the cooling system for six reactors, which caused significant pollution with radioactive contaminated water.

Despite their reliability and weather independence, nuclear power stations are not very flexible in terms of the ability to vary their output to meet immediate changes in power demand. Nuclear power stations work most efficiently when they produce a steady level of output over a long period and are therefore more suited to contributing to the 'base-load' of electricity demand.

Nuclear power stations are also expensive and time-consuming to construct and to decommission at end of life. These issues, coupled with widely held opposition to nuclear power on safety grounds, have greatly hindered the further expansion of nuclear power in developed countries. For example, Italy has banned nuclear power stations and Germany has declared that its existing nuclear plants will close by 2022. Nevertheless, some governments remain committed to nuclear power, e.g. over 80% of electricity in France is nuclear generated and China has an ongoing and substantial programme for constructing new nuclear power stations.

Combined Heat and Power (CHP)

Combined Heat and Power (CHP) is the generation of usable heat and power (usually electricity) in a single process.

CHP systems can be employed over a wide range of sizes, applications, fuels and technologies. In its simplest form, CHP employs a gas turbine, engine or steam turbine to drive an alternator and the resulting electricity can be used either wholly or partially on site, with any excess being supplied to the national grid system. The heat produced during power generation is recovered, usually in a heat recovery boiler, and can be used to raise steam for a number of industrial processes, to provide hot water for space heating or, with appropriate equipment installed, cooling.

Because CHP systems make extensive use of the heat produced during the electricity generation process, they can achieve overall efficiencies in excess of 70% at the point of use. In contrast, the efficiency of conventional coal-fired and gas-fired power stations, which discard this heat, is typically around 38% and 48% respectively. Efficiency at the point of use is lower still because of the losses that occur during transmission and distribution.

CHP is a form of decentralised energy technology. CHP systems are typically installed on site, supplying customers with heat and power directly at the point of use and therefore helping to avoid the significant losses which occur in transmitting electricity from large centralised plant to the customer.

Biodigesters

As the name implies, biodigesters use bacteria to break down organic matter (such as pig manure). Anaerobic conditions are maintained in these digesters, the action of the bacteria producing biogas, composed mainly of methane and CO_2 but with some hydrogen sulphide. After purification, the methane can be used as a fuel, in the same way that conventional domestic gas is used. Depending on the sophistication of the biodigester (there are numerous commercial designs), other by-products of the process that can be extracted include recycled water (but not suitable for drinking!) and high-grade fertiliser.

> **DEFINITION**
>
> **ANAEROBIC CONDITIONS**
>
> Low oxygen levels.

Methane Recovery

Biodigesters, discussed above, are a form of 'methane recovery' in the wider sense but this is a term most commonly associated with landfill sites. Much of the waste sent to landfill is organic and biodegradable. The action of bacteria under the anaerobic conditions in landfill sites produces 'landfill gas', which is largely methane (around 50-60%). Methane is a potent greenhouse gas and also flammable. It can therefore:

- Contribute to climate change.
- Present a risk of explosion either at the landfill site itself or at local residential areas (as it seeps through the ground laterally and vertically).

However, if it is collected, it can be used directly for fuel and to generate electricity in 'gas-to-energy' projects (using engines/turbines). The recovered energy can either be used for running the operations at the landfill site alone or for the local community, or exported.

The landfill gas is typically collected by sinking a series of vertical wells into the landfill site. These wells are connected via pipelines and the gas is drawn off (with the aid of a vacuum system) to a collection point for purification (using a scrubber) and processing (such as compression, combustion in an engine generator and hence conversion to electricity).

Biomass

In addition to wood, a number of agricultural crops are increasingly being grown specifically to generate biomass for use as fuel, including special grasses (Miscanthus) and hemp. Simple sugars from crops, such as sugar cane and sugar beet, can also be fermented to produce ethanol, which can be used either directly as a fuel or as an additive in petrol, e.g. bioethanol from the processing of sugar cane is a common vehicle fuel in Brazil.

Vegetable oils from crops, such as rapeseed, can also be converted to biodiesel. The use of land for growing energy crops rather than food, however, is controversial, especially in developing countries that may experience food shortages. As with any agricultural crop, high inputs of artificial fertilisers may also be used to generate high yields and may result in negative environmental impacts - especially diffuse pollution of watercourses with nitrates.

9.2 Renewable Sources of Energy

Benefits and Limitations of the Use of Alternative Energy Sources

Alternative energy sources clearly have many benefits to offer, many of which they have in common. However, they also have limitations.

> **TOPIC FOCUS**
>
> **Benefits and Limitations of Alternative Energy Sources**
>
> All alternative energy sources have the benefit of:
>
> - Reduced or zero CO_2 production and therefore:
>   - Reduced adverse impact on the environment.
>   - Less contribution to climate change.
> - Being 'renewable' or 'comparatively renewable' (except nuclear).
>
> Other benefits and limitations are summarised in the following table:
>
> | Alternative Energy Source | Benefits | Limitations |
> | --- | --- | --- |
> | Solar | <ul><li>Remote areas.</li><li>Close to where energy is required.</li><li>No emissions.</li></ul> | <ul><li>Unable to control how much and when.</li><li>No power generation at night.</li></ul> |
> | Wind | <ul><li>No emissions.</li><li>Remote areas.</li><li>Free form of motive power.</li><li>Small-scale operation as a local source of energy.</li><li>Plant can be prefabricated off site.</li></ul> | <ul><li>Unable to control how much and when.</li><li>Only generates power when there is wind.</li><li>Noise generated by turbines.</li><li>Loss of visual amenity.</li><li>Construction and maintenance costs can be significant.</li><li>Have to be large to provide sufficient energy for large-scale demand.</li><li>Remote from demand means that long supply cables required with subsequent energy transmission loss.</li><li>Objections by some to turbines.</li></ul> |
> | Hydroelectric | <ul><li>Dams and reservoirs provide additional recreational resources.</li><li>Long useful life of plant.</li></ul> | <ul><li>Construction and loss of habitat (e.g. by flooding valleys).</li><li>Reservoirs can generate methane from anaerobic decomposition (tropical regions).</li></ul> |
>
> *(Continued)*

Renewable Sources of Energy — 9.2

TOPIC FOCUS

Alternative Energy Source	Benefits	Limitations
Wave and Tidal Power	• No emissions. • No waste products. • Limited running costs.	• Unable to control how much. • Only produce power when there is wave or tidal action. • Ensuring the generator and associated equipment remain anchored in place.
Geothermal	• No emissions to air. • Remote locations. • Reliable fuel source.	• Often relatively large amounts of land required.
Nuclear	• Relatively small amounts of fuel produce a lot of power. • Output is reliable and not weather-dependent. • Very low emissions of CO_2 and other greenhouse gases.	• Generates significant volumes of potentially dangerous radioactive wastes. • Expensive and time-consuming to construct and decommission. • Inflexible in terms of the ability to 'turn on and off' in response to changing power demands. • High-profile accidents have created a significant degree of public and political opposition.
Combined Heat and Power (CHP)	• Efficient method for utilising both fossil and renewable fuels. • Highly efficient. • Usually generated at the point of use.	• Emissions to air. • Loss of visual amenity. • Transport of fuels to site.
Biodigesters	• Organic matter decomposes naturally — biodigesters utilise a waste product. • Relatively low cost.	• Possible odour problems. • CO_2 is still produced when methane is burnt. This, although less potent, is still a greenhouse gas.
Methane Recovery	• Methane is generated by landfill sites as part of the decomposition process. • Reliable source of fuel from a landfill as it is produced for many years even after closure. • Low cost.	• Uncollected methane is a greenhouse gas and presents a risk of explosion. • CO_2 is still produced when methane is burnt. This, although less potent, is still a greenhouse gas.
Biomass	• Renewable resource. • Biomass crops can be grown widely in many locations.	• Uses land that could be used for growing food. • Indirect pollution associated with high inputs of artificial fertilisers.

9.2 Renewable Sources of Energy

Energy Supply in Remote Locations and Developing Countries

Developing countries face an important challenge in making modern energy services available to help alleviate extreme poverty and meet other societal development goals. However, pollution from energy generation in developing countries is growing rapidly and makes a significant contribution to issues, such as poor air quality and climate change, presenting a serious risk to the environment and human health. Efficiency measures are not considered in the rush to provide energy to develop.

On-Site Energy Generation and Storage

Microgeneration involves the small-scale generation of heat and electricity in homes and businesses. This occurs as an alternative to taking power from a centralised grid system. Often this approach uses technologies that are more sustainable such as solar power or combined heat and power. Batteries are often used for some microgeneration techniques where energy generation is variable, such as solar power; energy can then be stored for use when it is needed. As power is created locally, transmission lines are shorter, resulting in minimal loss in power during transmission when compared to a centralised grid system.

Benefits and Limitations of Using Emerging Technologies for Energy

There are many emerging energy technologies that could lead to a more sustainable way of generating energy. These include such technologies as:

- Marine solar – sea-based solar panels.
- Static compensators - enables voltage stability.
- Molten salt reactors - nuclear fission where the nuclear reactor coolant and/or fuel is a molten salt mixture.
- Green hydrogen - hydrogen that has been generated using renewable energy.

The benefits and limitations of these emerging technologies vary by individual technology. However, general benefits and limitations of the group include:

- **Benefits**
    - Environmental impact – emerging technologies will reduce carbon dioxide and other emissions to the environment.
    - Energy security - countries will generally be more self-sufficient by adopting new technologies.
    - Fuel poverty – the potential generating costs will eventually be low.
- **Limitations**
    - Cost - installation costs can be high.
    - Technical issues - installation is more complex than standard installations (e.g. a boiler).
    - Lack of infrastructure - particularly relevant to hydrogen.
    - Novel technologies - these technologies have yet to be proven on a large scale.

STUDY QUESTIONS

2. Outline the three main ways in which we use solar power.
3. What is Combined Heat and Power (CHP)?

(Suggested Answers are at the end.)

Energy Efficiency

IN THIS SECTION...

- Energy efficiency can achieve both reductions in greenhouse gas emissions and financial savings in energy bills.
- Reducing emissions of CO_2 and other greenhouse gases is essential to prevent or reduce the effects of climate change.
- A key part of energy management is energy monitoring.
- There are a number of key measures to improve energy efficiency, including:
    - Insulation.
    - Choice of energy-efficient equipment and vehicles.
    - Maintenance and control of systems.
    - Building design.
    - Planning peak load management.
    - Optimisation of vehicle use.

Benefits of Energy Efficiency

There is now a strong focus on the need to reduce energy consumption as well as develop new ways to provide it. Clearly, if we do not use as much energy, then less needs to be generated and therefore the environmental impact is reduced. Even the renewable energy supplies discussed above have some environmental impact from the:

- Production of the equipment.
- Need to travel to maintain it.
- Energy used in these activities.

It should therefore be clear that improving energy efficiency should be at the top of any hierarchy of controls. Generally, if energy is used more efficiently, it will lead to a reduced environmental impact. For example, reducing emissions of CO_2 and other greenhouse gases is essential if the detrimental effects of climate change are to be prevented or reduced. Numerous other environmental benefits will also accrue, such as reduction in the release of acidic gases such as oxides of nitrogen, in addition to improved air quality. The benefits of energy efficiency, however, also occur in other ways such as reduction in cost (less electricity is consumed and less carbon taxes will need to be paid) and an improved reputation of the organisation.

Energy Monitoring

A key part of energy management is energy monitoring. This will involve a continuous strategy of collecting and analysing consumption data, such as from meter readings. Ongoing monitoring may identify both improvements and areas of weakness. There should be a plan to undertake regular, detailed monitoring. How frequent this is depends on the nature of the organisation - monitoring can be straightforward or involve complex analysis.

Energy monitoring can:

- Understand the reasons for excessive energy use.
- Detect times when energy use is higher or lower than expected.
- Provide a visualisation of trends in consumption.
- Assist in forecasting future energy usage and costs when planning business changes.
- Diagnose specific parts of the organisation which are wasting energy.
- Quantitatively understand the impact of improvements implemented to reduce energy.

9.3 Energy Efficiency

Control Measures Available to Increase Energy Efficiency

Reducing Energy Use

There are a number of control measures available to manage and reduce energy use. None of these is likely to provide a full solution to the issue of energy management within an organisation. There must be an objective assessment of where energy is being used and/or lost or used inefficiently, so that a suitable combination of controls can be implemented and any spending required can be targeted effectively. Such a system of energy management can be found in **ISO 50001:2018** Energy management systems, which sets standards for measurement, documenting, reporting, design/procurement practices for systems and people that contribute to energy performance.

Insulation

Effective insulation is a medium-cost measure that can usually be implemented relatively easily. Common insulation techniques are loft- or roof-space insulation and cavity wall insulation. These can be as effective on commercial premises as they are on domestic homes.

The following image and table show the difference that good quality insulation can make in reducing heat losses from an ordinary house. In the thermographic image, the dark areas show the least amount of heat loss, while the white areas show the most heat loss.

Thermographic image showing heat loss from a house

Difference in heat loss between insulated and uninsulated house

Route	% Heat Loss from Uninsulated House	% Heat Loss from Insulated House
Roof	25%	5%
Walls	35%	10%
Doors	10%	5%
Floor	10%	5%
Windows	20%	10%

There are many EU Directives that cover energy usage and performance, e.g. **Directive 2010/31/EU on the Energy Performance of Buildings**, the principal objectives of which are to promote the:

- Improvement of the energy performance of buildings within the EU through cost-effective measures.
- Convergence of building standards towards those of member states which already have ambitious levels.

> **MORE...**
> More information on practical steps that can be taken by organisations to improve energy efficiency is available from the following source:
>
> https://www.carbontrust.com/resources/office-energy-efficiency-guides

Equipment Selection

Choosing the right equipment is likely to have a major impact on energy consumption. For instance, electric motors must be matched to the demand required of them:

- Buying one that is too small will mean it is running at full capacity most of the time and is likely to require more current.

- One that is over-sized will be running at too low a capacity and therefore not running efficiently.

If lighting systems are to be changed or installed in new buildings, designing the system correctly and installing the right equipment can ensure significant savings over the life of the building. The common causes of wasted energy with regard to lighting systems are:

- Lights being used unnecessarily.
- Lighting unoccupied buildings or rooms.
- Using lights when daylight provides sufficient light levels.

Significant savings can be made through the use of low energy lighting. Taking the standard tungsten halogen bulb as 100%, the table below shows potential savings:

Typical lamp efficacies

Lamp Type	Efficacy Lumens/Watt
Incandescent -- tungsten filament	6-14
Incandescent -- tungsten halogen	13-26
Compact fluorescent	45-70
Tubular fluorescent	38-106
Low pressure sodium	100-168
High pressure sodium	70-150
Metal halide	70-107
Light Emitting Diodes (LEDs)	25-100 Variable depending on colour and increasing as technology develops

Adapted from Display Lighting, Carbon Trust, 2012

Maintenance

A system will only remain operating efficiently if it is properly maintained. Whether it is servicing heating boilers, or ensuring lights are cleaned regularly, the same principles apply. An effective planned preventive maintenance system will help avoid expensive and inconvenient breakdowns and should, over time, reduce the operating costs as the equipment is kept in the most efficient operating condition.

Heating

Unless heating equipment is very old and unreliable, consideration should be given to improving the control of existing systems before the purchase of new equipment is considered.

There are three main types of controls for heating systems:

- **Simple switches** - may be manual or automatic and simply switch equipment on or off. Automatic switches can be used to preset the time at which the heating will turn on and off.
- **Complex time switches** - will often be able to compare indoor and outdoor temperatures and be set to bring a building up to a certain temperature by a specific time. This is especially useful where buildings such as offices are not occupied overnight but occupants do not want to come in to a cold building. With correct control systems, the start time can be varied depending on outside conditions.
- **Continuous controllers** - these will constantly monitor and adjust the system to maintain an even temperature within defined limits. They are likely to be more complex to set up, but also the most efficient system.

9.3 Energy Efficiency

A modern Building Energy Management System (BEMS) can significantly reduce the energy used to heat and cool a building. It also has the advantage that it is usually controlled centrally and so is less likely to be misused with temperatures set too high or too low. A centralised system also allows a more consistent temperature throughout a building, preventing the situation where different sections of a system are conflicting with each other to maintain a set point temperature.

Lighting

Lighting systems account for a significant amount of energy used in most buildings, and the fitting of suitable control systems can play a major role in reducing this energy consumption. There are two main categories of lighting controls:

Illuminated buildings will have significant energy use

- **Manual** - the simple on/off switch we are all familiar with. Proper labelling allows people to know which lights can be switched off when not required. Combined with the correct layout, so the layout of the switches matches that of the lights, this means people are more likely to switch off those lights not required. Significant savings are possible if lights are switched in rows running parallel to the windows and lights nearest the windows can be switched off separately from those further away.

- **Automatic** - controls such as timers and sensors can be used to automatically turn off lights when not required. Sensors can be for movement or light levels but the best systems combine the two. This way, even when people are in the room, if light levels are sufficient, the lights will not be switched on. It is important that automatic controls are set correctly to ensure that people are not left in darkness; this involves correct positioning of the sensors, correct sensitivity of sensors and correct timing before lights are switched off. This is especially important where people may not be moving around the room all the time and it may be necessary to have a manual override available.

Occupancy sensors can also be used in other applications, such as urinal flush control. The system would be able to control lighting, ventilation and flushing relative to occupancy and therefore produce savings in energy and water.

Building and Site Design

Considering energy efficiency at the design stage is key in making buildings more energy-efficient. For example, natural ventilation (rather than ventilation forced by a fan) can lead to considerable energy savings. Fresh air is required in buildings to provide oxygen, prevent odours and increase thermal comfort. Natural ventilation can be achieved in numerous ways, such as using openable windows or trickle vents. More complex methods include designing buildings to allow warm air to move upwards to upper openings; this will force cool air to be drawn in from outside. In very warm climates, natural ventilation may not always be possible and more conventional air-conditioning systems will be needed.

Optimising the size of the building for the number of staff who work there is another option - the less wasted space there is to heat and light then the less energy usage the building will have (hot desking for some staff members could help partly achieve this where feasible).

A further design option is to provide passive solar heating, which uses windows, floors and walls to collect and distribute heat during cool periods and reject heat during hot periods. This is a passive system as it does not involve any mechanical or electrical equipment. It takes advantage of the climate in the location of the building. Elements often considered in such systems in temperate regions include:

- Windows facing the midday sun in winter and being shaded in the summer.
- Reducing windows on other sides.
- Using suitable insulation to reduce seasonal excessive heat loss and gain.
- Using thermal mass (the ability of the building to soak up heat before it reaches the interior during the day and release it at night).

Energy Efficiency | 9.3

Planning Peak Load Management

Peak load management is a process often used by utility suppliers to manage the peak load required. Any user of electricity can use peak load management in conjunction with their electricity provider to assist with the management of power supplies, although it is most commonly used by heavy industry and large commercial users.

Many industrial users of electricity will sign up to an agreement to reduce power demands during peak periods. The customer gets a reduced price overall but pays a large penalty if they fail to reduce power usage when required to do so by the utility company.

> **DEFINITION**
>
> **PEAK LOAD/PEAK DEMAND**
>
> Terms used interchangeably to denote the maximum power requirement of a system at a given time, or the amount of power required to supply customers at times when need is greatest.

Fuel Choice for Transport and the Optimisation of Vehicle Use

For many organisations, transport represents a significant proportion of their energy consumption. This is especially so if the travel for staff getting to and from work is taken into account. There are now a number of different fuel sources available for transport and here we take a look at some of the more common ones as well as other ways to reduce energy use associated with transport:

- **Petrol and diesel** - still the most common transport fuel for most people and goods; produced from oil and very commonly available. The burning of petrol and diesel produces oxides of sulphur and nitrogen along with carbon monoxide, CO_2 and VOCs, and has been identified as one of the most significant contributors to climate change and air pollution. Diesel has an advantage, in that it is generally more efficient than petrol, resulting in reduced fuel consumption for comparable vehicles. However, it also produces more particulates and sooty emissions than petrol so is often seen to be 'dirtier' than petrol.

- **LPG** - the use of LPG (a mixture of propane and butane produced as a by-product of oil and gas production) has seen a steady increase around the world. Most cars which have been converted to run on LPG can also run on petrol; this is necessary due to the limited number of fuel stations providing LPG. It is not as efficient as petrol in that some loss of power is often found in comparison to the same engine running on petrol. However, it is currently around half the price of petrol and the emissions are significantly cleaner than those from a similar petrol engine, with fewer particulates and some 15% reduced CO_2 emissions.

- **Electric** - all-electric vehicles have also seen an increase in popularity. The range of electrically-powered cars, for example, is now over 300 miles (480 kilometres) and top speeds rival those of petrol- or diesel-powered vehicles. As technology improves it seems that it will only be a relatively short period of time before electric cars are equal to petrol and diesel cars in performance and practicality. The environmental benefits of electric vehicles are significant; they produce zero emissions in the location where they are used and indirect emissions from electricity generation can vary, but will be minimal if renewable forms of energy are the source of electricity. Electric vehicles also have widespread commercial uses.

- **Hybrid power** - these vehicles use a combination of an internal combustion engine (usually petrol) linked to an electric motor. The electric motor assists the petrol engine, thereby allowing a smaller petrol engine to be used and reducing fuel consumption; at the same time, the batteries for the electric motor are recharged by the petrol engine. At very low speeds, the electric engine can power the vehicle completely for short distances.

- **Biofuels** - championed as the saviour of the environment for many years, the benefits of biofuels are now being questioned by many environmental organisations. Produced from crops such as rapeseed and palm oil, it is now clear that first-generation biofuels (those produced directly from crops as opposed to the use of second-hand oils such as used cooking oil) are not CO_2 neutral and may be more harmful than ordinary oil products. There is great concern regarding the diversion of crops away from the production of food to the production of vehicle fuels, thereby adding to food shortages in many parts of the developing world.

9.3 Energy Efficiency

New Vehicle Technologies

Research into new technology to ensure vehicles have a reduced impact on the environment is occurring at a rapid pace. Current innovations include:

- **Advanced materials** - these are essential in boosting the fuel economy of vehicles as it takes less energy to accelerate a lighter object when compared to a heavier one. A 10% reduction in vehicle weight equates to a 6% - 8% improvement in fuel economy. Replacing iron and steel components with lighter weight materials such as aluminium, carbon fibre and polymer composites can significantly reduce vehicle weight.
- **Self-driving vehicles** - the use of self-driving vehicles is likely to reduce fuel consumption in various ways. On motorways, automated vehicles can drive extremely close to each other at high speed, so reducing aerodynamic drag. Automated vehicles can also be made to run on an eco setting that reduces fuel consumption considerably. With just about all cars being automated, the crash risk reduces dramatically (around 90% of traffic fatalities arise from human error). Such a vehicle's weight could be reduced substantially as the risk of an accident is very low.
- **Adaptive wheels** - a micro compressor can adjust the tyre pressure, the width of the rim and alter the amount of rubber in contact with the road to best suit road conditions. For example, the tyre pressure can be increased and the wheel made narrower for travel on dry roads.
- **Smart infrastructure** - information from road sensors can be used to assess the optimum speed a car should travel at. Smart street lights can provide information on the volume of traffic and control light sequencing to increase flow and as such reduce emissions.

Reduction and Optimisation of Travel

Other ways to reduce energy consumption related to travel are often termed 'Green Travel Plans'. These plans involve an organisation looking at how employees travel to, from and during work and ways to reduce the dependence on cars. This could be by:

- Working with local bus companies to change routes or timetables.
- Providing incentives, for example:
    - Interest-free loans for rail season tickets.
    - The 'cycle to work' scheme supported by some governments, where employers can offer interest-free loans to purchase a bicycle to get to and from work. This allows an employee to either purchase a better bicycle or spend less on the same bicycle they would have bought.
- Employee car-sharing schemes - experience has shown that it is quite difficult to get people to initially sign up for such schemes but, once signed up, they use them enthusiastically. It is noteworthy that when fuel prices increase dramatically, so does the popularity of these schemes.
- Training staff in efficient driving, for example:
    - Under-inflated tyres and roof racks increase fuel consumption.
    - Avoiding sharp acceleration and braking reduces fuel consumption.
    - Planning journeys to reduce the amount of time spent in congestion will reduce fuel consumption.
- Use of technologies such as video conferencing, which may eliminate the need for travel.

Energy Efficiency | **9.3**

TOPIC FOCUS

In summary, the key measures an organisation can introduce to **reduce energy consumption associated with transport** include:

- Choice of appropriate fuels.
- Car sharing.
- Financial incentives to cycle to work or use alternative transport modes.
- Planning travel outside peak times.
- Training staff in efficient driving behaviour.
- Working with local bus companies to change routes or times of buses.
- Use of technology such as teleconferencing rather than travel.

STUDY QUESTION

4. Explain some of the reasons why energy efficiency initiatives fail.

(Suggested Answer is at the end.)

Summary

This element has dealt with the sources of energy and energy efficiency.

In particular, this element has:

- Outlined how fossil fuels (coal, oil and gas) are formed from the organic remains of marine micro-organisms (oil and gas), and land-based vegetation (coal).
- Explained the benefits of the use of fossil fuels as an energy source, including: a straightforward combustion process; relatively inexpensive; easily transported, etc.
- Highlighted the limitations of fossil fuels: major contributors to climate change, cause acid rain, use a non-renewable source of energy, prices fluctuate according to world politics, etc.
- Discussed other sources of energy including: solar, wind, wave and tidal power, geothermal, nuclear, Combined Heat and Power (CHP), biodigesters, methane recovery and biomass.
- Outlined the problems associated with energy generation and supply in developing countries and remote regions.
- Explained that the benefits of alternative energy sources include reduced or zero CO_2 production, and the possibility of generating power in remote areas or close to where power is required. Generally, they are cleaner systems with few emissions at the point of generation.
- Shown that renewable sources (e.g. wind, wave, tidal and solar power) have a limitation in that they are out of our control in terms of how much power they will generate and when. Disadvantages of nuclear power are problems of acceptance, of site location and long-term secure storage requirements for waste materials.
- Explained that a significant result of energy efficiency will be a reduction in CO_2 emissions. In addition, savings in energy bills can be achieved by using peak load management (particularly important for industrial users).
- Discussed control measures available to manage and reduce energy use, including:
    - Insulation to prevent heat loss.
    - Choice of equipment: type of fuel used, efficiency of the equipment.
    - Maintenance: to ensure equipment remains efficient.
    - Adequate equipment control systems.
    - Building design.
- Outlined how organisations should review their transport use (travel to work and for work purposes) and ensure they choose from the various fuel sources available with protection of the environment in mind.

Exam Skills

Question

> **Scenario**
>
> You have been tasked to reduce energy consumption in the large office in which you work. Your manager is particularly interested in reducing the financial cost of energy. Currently few energy efficiency measures have been implemented although energy consumption is monitored every month.
>
> **Task**
>
> What measures can be implemented to reduce the financial cost of energy at the organisation?
>
> **(8 marks)**

Approaching the Question

Think about the steps you would take to answer the question:

- Read the scenario carefully. With this question you need to develop some reasons as to measures to reduce the financial cost of energy. Note: the question states 'financial cost' so measures may not necessarily reduce consumption (although most are likely to reduce consumption and reduce financial costs).
- Now look at the task – prepare notes on:
    - Energy efficiency measures for a large office.
    - Other measures to reduce the financial cost of energy in a large office.
- Consider the marks available. In this case, there are 8 marks available so you should provide 8 measures to reduce the financial cost of energy in a large office.
- Read the scenario and task again to make sure you understand them and have a clear understanding of ways to reduce the financial cost of energy in a large office.
- Jot down an outline plan - this might include:
    - Design, maintenance, lighting, switch-off scheme, alternative sources, peak load management, transportation, insulation, energy management.

Now have a go at the question yourself.

Exam Skills

Example of How the Question Could be Answered

The initial design and choice of equipment for the task is important. Purchasing equipment with reduced energy consumption in comparison with present inefficient equipment will reduce costs in the long term.

Equipment will also need to be maintained so that it continues to work effectively. Electrical costs can be minimised by lighting controls, such as fitting motion sensors in corridors or other relevant areas.

Compact fluorescent lights and other more efficient light types could also be fitted. The use of high-energy-rated equipment will also reduce consumption and cost. Lights that have a high energy rating should be purchased by the organisation.

The organisation could also implement a switch-off scheme - this might include provision of labels, posters and basic training to staff to switch lights and equipment off when not in use.

The use of alternative forms of energy should also be considered. Combined heat and power will, for example, make use of waste heat from power generation which could be used to heat a building.

Peak load management could also be used as this will ensure that an organisation pays less for its energy if it uses less at peak times.

Costs could also be saved through improvement in transportation with fuel type, effective route planning and training of drivers all leading to potential cost reductions.

Insulation of pipework and buildings and not heating areas that are unoccupied can also lead to cost reductions.

Employing an energy management company to look at energy consumption for a large site, including supplier and tariff analysis, can also significantly reduce costs associated with energy.

Unit EMC1

Final Reminders

Now that you have worked your way through the course material, this section contains some reminders to help you prepare for your NEBOSH open-book exam. It summarises the advice on how to approach your revision and the exam itself and has some hints and tips.

Unit EMC1: Final Reminders

Preparing for the Exam

Open-book exams require advance planning. As you work through your studies, it's important to familiarise yourself with the study materials so that, at the time of the exam, you can find what you need quickly. RRC's course materials cover all the syllabus topics but to ensure that you do well in the exam, we recommend doing some additional reading. The study text provides some useful links to external sources - have a look at the 'More...' boxes within the materials, these contain useful links to relevant topics.

> **MORE...**
>
> Further information on the science behind climate change can be found in the Intergovernmental Panel on Climate Change (IPCC) Fifth assessment report - climate change 2013: The physical science basis, available at:
>
> www.ipcc.ch/report/ar5/wg1

'More...' boxes provide relevant links to further reading material (taken from RRC's EMC1 Study Text)

At the time of the exam, you should not be reading information from your course materials for the first time or even re-reading the study text, you will simply run out of time. Being familiar with the materials will give you more time to concentrate on the scenario and less on frantic searching!

Don't forget that the normal requirements of an invigilated exam don't apply, so you can highlight and annotate your materials to help you locate topics easily and use your notes on the day. The Open University webpage has some great tips on highlighting and annotating materials for revision purposes, and can be accessed at: https://help.open.ac.uk/highlighting-and-annotating.

Keep your notes organised in advance; by doing so, you will be able to easily identify the relevant parts to compose your answers. This will ensure you optimise your time during the exam.

Revising for the Exam

One of the most common misconceptions about open-book exams is that there's no need to revise for them. In fact, you should study for them just as you would for any other exam! You won't be asked to recall information in the same way as for a closed-book exam but you still need the knowledge in order to apply it effectively and you need to be able to demonstrate that you have met the learning outcomes. Remember, the exam presents you with a problem in the form of a scenario to which you will need to give a solution, so you will need to use your knowledge and apply it to solve the problem.

Revision Tips

Using the RRC Course Material

Read through all of the topics multiple times. This might be done by skimming over all of the content of Unit 1 to get a feel for structure and topics, followed by a more thorough read-through that jumps over the most complex topic areas, then a detailed read where you attempt to crack the complex topics.

Remember that understanding the information, and being able to remember and recall it, are two different skills. As you read the course material, you should understand it. In the exam, you have to be able to remember, recall and apply it. To do this successfully, most people have to go back over the material repeatedly.

Check your basic knowledge of the content of each element by reading the element Summary. The Summary should help you recall the ideas contained in the text. If it does not, then you may need to re-visit the appropriate sections of the element.

Unit EMC1: Final Reminders

Using the Syllabus Guide

Download a copy of the NEBOSH Guide to the course, which contains the syllabus, from the NEBOSH website.

Map your level of knowledge and recall against the syllabus guide. Look at the content listed for each element in the guide. Ask yourself the following question:

'If there is a question in the exam about that topic, could I answer it?'

You can even score your current level of knowledge for each topic in Unit 1 of the syllabus guide and then use your scores as an indication of your personal strengths and weaknesses. For example, if you scored yourself 5 out of 5 for a topic in Element 1, then obviously you don't have much work to do on that subject as you approach the exam. But if you scored yourself 2 out of 5 for a topic in Element 3 then you have identified an area of weakness. Having identified your strengths and weaknesses in this way, you can use this information to decide on the topic areas that you need to concentrate on as you revise for the exam.

You can also annotate or highlight sections of the text that you think are important.

Another way of using the syllabus guide is as an active revision aid:

Pick a topic at random from any of the Unit 1 elements.

Write down as many facts and ideas that you can recall that are relevant to that particular topic. Go back to your course material and see what you missed, and fill in the missing areas.

Setting Up for the Exam Day

It is recommended that you study in the same room and environment where you will carry out the exam, to ensure you are comfortable and set for the exam.

There are some things you can do to ensure you have the best possible set-up for the day:

- Make sure you can sit comfortably so that you are not distracted by uncomfortable posture. Ensure good lighting and a comfortable temperature.

- Know where your study materials are so that you spend less time looking for them at the time of the exam. Have your study materials within easy reach.

- If you live with anyone, make sure they are aware of when you are taking an exam to avoid unnecessary interruptions and distractions. Placing a friendly sign on your door may be a useful reminder for them!

- Switch off your phone, television and any other devices that may distract you.

- Have water and snacks handy.

- Ensure you can keep your computer or other device charged up.

- If you can't take the exam at home, book a quiet room with good lighting, charging point and Internet connection.

Unit EMC1: Final Reminders

What to Do on the Day

On the day of the exam, you will be able to access the exam from 11.00 am UK time by logging in to the NEBOSH platform and downloading the file. You will have 24 hours to do the exam, starting from when the exam paper becomes available. This does not mean it should take you 24 hours to do the exam, nor does it mean that you have to be working for all that time; the 24-hour window is designed to allow time for you to read and analyse the exam questions, access your course materials, plan your answers, complete and submit the assessment, as well as take necessary breaks and fulfill your other everyday commitments. The paper should take around 4 to 5 hours to complete so make sure you are aware of the time.

So, how do you best utilise this time?

This is when your planning, studying and hard work will pay off. You will have your materials ready so you will be set up for a strong start.

You are not expected to write more than the current 3,000 words in total. You are allowed a 10% margin – you will not gain marks for going beyond this, so your answers should be relevant, clear and focused. RRC would strongly advise that you **do not write more than 3,300 words** in case examiners choose not to read beyond 3,300 words and you therefore miss out on marks.

Use your time wisely: work at your own pace but don't leave everything until the last minute. Review your materials, draft up your answers and allow time to make amendments. Take time to read the exam questions carefully. Refer to your prepared materials and notes with the following in mind:

- Your work should be your own, in other words do not copy content without referencing the source or this counts as plagiarism (more on plagiarism later).
- Do not communicate with anyone about the assessment.
- Do not ask or allow anyone to proof-read or help you with your work.

Another common mistake when doing an open-book exam is to refer to as many materials as possible; don't fall into this trap, you must be selective! Use only the materials that you need - again, this is why preparing them in advance is so important!

Don't become over-reliant on materials either, you must apply your own knowledge and argument. You want the materials to support your answer, not take over!

For details on how to download and submit your exam, please read NEBOSH's Technical Learner Guide on the NEBOSH website.

Approaching the Exam Questions

The open-book exam will test you on your ability to "demonstrate analytical, evaluation and creative skills as well as critical thinking" and how you apply your learning to your answers. In other words, you will need to show what you can do with your knowledge to solve the problems presented to you – and this may take practice. To increase your chances of success and improve your confidence, we strongly advise that you complete a mock exam and get feedback from your tutor to prepare you for the real exam.

Plagiarism and Malpractice

You should follow the instructions and adhere to the guidance on the open-book exam. The answers that you submit must be your own. Any cases of suspected plagiarism will be investigated and any breaches will be dealt with in line with NEBOSH's malpractice policy which you can find on the NEBOSH website..

You must ensure that what you submit is your own work and if you quote or paraphrase anyone else's work, this must be referenced or it would constitute plagiarism.

The following counts as plagiarism:

- Inserting another author's sentences, paragraphs and ideas without referencing them, whether these are published or unpublished.

Unit EMC1: Final Reminders

- Paraphrasing another author's work without referencing them.
- Collaborating with someone else (e.g. another learner) and submitting work that is either identical or very similar to theirs while claiming it was your own work.
- Paying someone to complete the work for you and submitting it as your own.
- Impersonation - when you ask someone else to complete the work for you and you pass it off as your own.

Your open-book exam will be marked by a NEBOSH examiner and will be scrutinised for plagiarism.

When taking a non-invigilated open-book exam you will need to declare that your submission is your own work and that you have not received help from anyone else. You will need to confirm you have read, understood and abided by NEBOSH's rules, by signing a Declaration.

Please note that NEBOSH reserves the right to submit your assessment to a plagiarism detection software package.

References

NEBOSH Open Book Examinations: Learner Guide - Guidance for preparing for an open book examination, NEBOSH, 2021

(https://www.nebosh.org.uk/digital-assessments/certificate/resources-to-help-you-prepare/)

NEBOSH Environmental Management Certificate - Unit EMC1: Management and Control of Environmental Aspects, RRC Study Text, 2021

Note-taking techniques, The Open University, 2020

https://help.open.ac.uk/highlighting-and-annotating

NEBOSH Environmental Management Certificate - Qualification guide for learners, NEBOSH, 2021

(https://www.nebosh.org.uk/qualifications/environmental-management-certificate/#resources)

NEBOSH Open Book Examinations: Technical Learner Guide, NEBOSH, 2020

(https://www.nebosh.org.uk/digital-assessments/certificate/resources-to-help-you-prepare/)

NEBOSH Environmental Management Certificate - Unit EMC1 Mock Exam, RRC, 2021

Policy and Procedures for Suspected Malpractice in Examinations and Assessments, NEBOSH, 2019

(https://www.nebosh.org.uk/policies-and-procedures/malpractice-policy-and-procedures/)

Good luck!

Unit EMC2

Practical Assessment Guidance

The aim of this section is to help you prepare for the NEBOSH Environmental Management Certificate, Unit EMC2: Practical Assessment.

This guidance is heavily based on the Guidelines for Learners published on the NEBOSH website which you should download and read carefully.

Practical Assessment Guidance

Introduction to this Guidance

The practical assessment is broken down into different parts. These are presented sequentially so that you can gradually work your way through each part, adding to your understanding of the practical assessment as you go.

It is better to wait until you have studied each element and have looked at all parts of the Practical Assessment Guidance before you start.

The images shown are for illustrative purposes only. They are reproduced from the NEBOSH Guidance to Unit EMC2 and the official forms and are subject to change by NEBOSH.

Introduction to the Practical Assessment

The practical assessment requires you to carry out an environmental aspects and impacts assessment in your workplace. There are several parts to this assessment as shown in the diagram below:

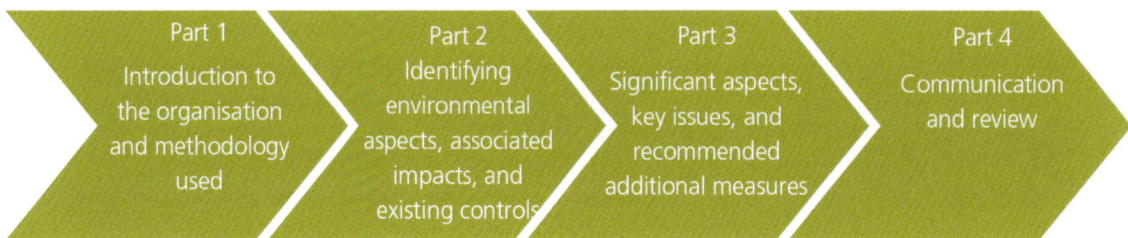

Part 1 — Introduction to the organisation and methodology used
Part 2 — Identifying environmental aspects, associated impacts, and existing controls
Part 3 — Significant aspects, key issues, and recommended additional measures
Part 4 — Communication and review

Once you have completed each part, you must write this down on the standard form supplied by NEBOSH in its online Assessment Pack. You can download this form from: www.nebosh.org.uk. It is important that you use this standard NEBOSH form for the assessment – **the use of non-standard forms might result in your assessment receiving a referral**. NEBOSH recommends that you complete the forms electronically; this has the added benefit of looking more professional.

Practical Assessment Guidance

Part 1 - Introduction to the Organisation and Methodology Used

Part 1 Introduction to the organisation and methodology used

Part 2 Identifying environmental aspects, associated impacts, and existing controls

Part 3 Significant aspects, key issues, and recommended additional measures

Part 4 Communication and review

The Part 1 Form

Part 1 of the NEBOSH form is shown below:

1. Introduction to the organisation and methodology used

You should aim to complete this section in 200 - 250 words

Name of organisation*	
Site location (town / area)	
Number of workers	
General description of the organisation	
Description of the area or process to be included in the assessment	
Any other relevant information	

* If you're concerned about confidentiality, you can invent a false name and location for your organisation. All other information provided must be factual.

Name _____ Learner number _____

You should aim to complete this section in 100 - 200 words

How the aspect and impact assessment was carried out, including: - sources of information consulted; - who you spoke to; and - how existing controls were identified	

Note: this section can be completed after you have competed your full assessment.

Note: These forms are for reference purposes only. Please visit the NEBOSH website to obtain the official forms to submit your assessment.

© RRC International — Unit EMC2 - Practical Assessment Guidance

Practical Assessment Guidance

Description of the Organisation

The first step is to provide a brief description of the workplace and write this down in the space provided on Part 1 of the NEBOSH form. **Make sure that you include all of the information that is requested at the top of this form**.

Ordinarily the workplace that you choose to describe would be the organisation that you work for but it doesn't have to be. It could be a location such as an office or you might choose to carry out this assessment at a larger, complex workplace.

You can provide a false name and/or location for your organisation if you like, to protect the identity of the organisation involved and maintain confidentiality. Everything else must be factual.

You need to think about the scope of the aspect and impact assessment exercise. You should choose an area that is large and interesting enough to offer a broad range of aspects and impacts. But you also want the whole exercise to be manageable in the time that is available to you. For example, you might look at the whole organisation or just a specific department or division. If you work for a large organisation, we recommend that you concentrate on one site.

If that site is very large and complex, then focus on one part of that site. Bear in mind that in a large organisation you do not need to assess every single aspect. You need to identify and assess a representative handful of the most significant aspects from the workplace. So being in a larger, more complex organisation is an advantage because you can selectively pick and choose your aspects from the broad range available. In a small, low-risk organisation, it might be difficult to get a representative handful of significant aspects.

So, in the space provided, include the organisation's name and location (even if they are fictitious). Include the number of workers and the shift patterns that are worked. Give a good general description of what the organisation does and the site layout so that the examiner can picture the kind of workplace, the products or services involved, and the sorts of activities that are likely to be carried out there. You could also identify the vulnerable receptors, such as rivers and housing that are in the vicinity of the site.

You do not need to write a long essay. But you must include a good description and you must address all of the key issues identified at the top of the form. Do not skip any of the core information even if it seems obvious or unimportant. The key thing is to give the examiner (who will be unfamiliar with your workplace) a quick overview of the location you have chosen for the assessment so that they can picture it. NEBOSH recommends that you write 200-250 words for the description.

Methodology Used

The second step requires you to explain how you carried out the aspects assessment (your methodology). Write this down in the space provided on Part 1 of the NEBOSH form. **Remember to include all of the information that is requested at the top of this form**.

You should include things such as the sources of information that you consulted, who you spoke to, and how the aspects and impacts and controls were identified. For example, if you looked at audit reports to identify some of the significant impacts, or you did a site walkabout, then write about these activities here. If you spent a lot of time researching information online, then say so. State which websites you looked at and which documents you used. If you talked to workers or managers, then say so. You can even give people's job titles, if they were key sources of information, but don't include personal names. The only personal name that should appear anywhere on the assessment is your name at the bottom of every page in the space provided.

NEBOSH recommends that you write between 100 and 200 words for this explanation but you might write more or less depending on the nature of your workplace and how you gathered your information. Again, do not skip the core information requested even if it seems obvious or unimportant.

Once you have filled in the relevant forms of Part 1, that's the first step completed.

Practical Assessment Guidance

Part 2 – Identifying Environmental Aspects, Associated Impacts, and Existing Controls

The second step in the practical assessment is to carry out an aspects and impacts assessment exercise and record the results on Part 2 of the NEBOSH form.

The Part 2 Form

Part 2 of the NEBOSH form is shown below:

2. Part 2 - Identifying environmental aspects, associated impacts, and existing controls

Activity, product, or service	Aspect	Operating condition(s)	Associated impacts	Existing controls	Significance rating	Justification for each Significance rating

Practical Assessment Guidance

Activity, product, or service	Aspect	Operating condition(s)	Associated impacts	Existing controls	Significance rating	Justification for each Significance rating

Note: These forms are for reference purposes only. Please visit the NEBOSH website to obtain the official forms to submit your assessment.

This table appears self-explanatory but it needs to be completed by carefully reading and following the NEBOSH guidance. We will start with the first step where you are required to identify the activity, product or service.

Please note that Part 2 of the aspects and impacts assessment form does not give any indication of word count. As this assessment is being done for the purposes of an academic assessment (rather than for work), you should write enough information so that the examiner can clearly understand your intention. Remember that the examiner does not know your workplace and will not be visiting it. It is better to explain things properly rather than write too little on the form.

Activities, Products or Services

You must include at least four different activities/products/services from your chosen site.

The activity, product or service selected should be large enough for meaningful examination and small enough to be sufficiently understood; e.g. an organisation may have many different compressors that use energy on-site, so depending on the size of the site, the energy use for compressors may be identified as one aspect.

A process-flow diagram is often developed to identify activities on a site (from goods in to despatch) and ancillary activities that are not part of the main process should not be forgotten (e.g. maintenance, fuel storage, office activities). A walk around the site or viewing a site plan will also be useful in determining activities.

Practical Assessment Guidance

Aspects, Operating Conditions and Impacts

You need to identify at least 10 different environmental aspects in your assessment with at least one associated impact (often there will be more than one impact and you should include all of them).

A simple way is to consider the inputs and outputs from an activity, product or service - these are the environmental aspects. Issues that are commonly addressed include:

- Emissions to air.
- Releases to water.
- Waste management.
- Contamination of land.
- Use of raw materials and natural resources.

Next you need to consider the operating conditions of each aspect. **To pass the assessment you must include two of the following conditions:**

- Normal: all planned activities need to be considered.
- Abnormal: non-routine such as maintenance and cleaning.
- Accidents/incidents/emergencies: reasonably foreseeable incidents should also be considered, e.g. fire or chemical/oil spillage.

The next stage is to determine environmental impacts of each aspect. These are the changes to the environment that are caused by the aspect.

As covered in Unit EMC1, aspects and impacts may be positive or negative (it is likely most you identify will be negative), direct, indirect or cumulative.

MORE...
https://blog.rrc.co.uk/2014/01/06/identifying-environmental/

Existing Controls

The next part of the form requires you to provide information on the current controls for each impact(s). You need to write a brief explanation of all of the current control measures that are in place.

Remember that some of the control measures will be physical things such as bund walls or abatement equipment. But also remember that many of the control measures will be administrative or procedural, such as a checklist for a particular maintenance task, worker training or maintenance programmes on items of equipment.

You need to include all of the current control measures that are relevant to the impact in question. This means that you might have to do some research to find out the full range of control measures. You might do this by talking to workers, talking to managers, looking at policy documents and records, and/or carrying out inspections in the relevant area. If you do carry out any of these activities, remember to write about them in the methodology section of the practical assessment in Part 1.

The key here is that you include enough information about the existing control measures so that the examiner can see what is being done and can make a judgment about their suitability. So you need to briefly describe what is being done, demonstrating that you are able to do a suitable assessment. Don't just write single words or very short phrases such as 'training' and don't simply refer to one control repeatedly as the only control measure.

Significance and Rating

Now you have identified the aspects and impacts, you need to determine which are significant. NEBOSH provides a significance rating table to assist in making this decision. **You must use this table to evaluate your aspects and impacts**.

Practical Assessment Guidance

Justification for significance rating	Significance rating
• Aspect is not currently controlled under normal condition • 'Abnormal' condition would breach regulations • Evidence of / high likelihood of immediate environmental damage • Severity of impacts is extensive • In breach of legislation • Interested parties have voiced concerns • Repeated complaints received (internal/external)	4 (Major) Immediate action and control required
• Aspect is not fully controlled under normal conditions • Severity of 'normal' condition impacts are moderate • Abnormal condition impacts are likely to occur and/or unlikely to be detected • Impacts are likely to increase based on planned activities • Emergency situation would cause major environmental impact • Potential breach of legislation • In breach of company policy or other non-legal compliance obligation • Financial / business threat from impacts • Interested parties aware of issue • Complaint received (internal/external) • Multiple instances of same issue / large volumes involved	3 (Significant) Action required within set timeframe, with scheduled review of effectiveness
• Aspect controlled under normal conditions • Abnormal condition impacts are unlikely to occur and/or likely to be detected • Severity and likelihood of any impact is low • No potential for breach of legislation or other compliance obligation • Single instance of issue / small quantities involved	2 (Minor) Additional actions possible; recommend to review in the future
• Aspect controlled under normal and abnormal conditions • No additional impact arising from emergency condition • Severity and likelihood of any impact is minimal • Very small quantities involved	1 (Negligible) Limited additional action possible; review in the future may be required

Note: meeting one of the criteria is enough for the aspect to be considered at the highest rating.

Note: This rating table is for reference purposes only. Please visit the NEBOSH website to obtain the official forms that you must use to complete your assessment.

For each aspect you need to choose the significance rating that best reflects the aspect. Your aspect only needs to meet one criteria in a category for it to be considered at the highest rating. Place in the 'Significance rating' column on the form the score and description of the significance rating, for example that would be 4 (Major) or 2 (Minor).

In the criteria column you need to copy across from the significance rating table the reason why you have chosen the rating from the bullet-pointed criteria list, so this might be for 3 (Significant) 'Potential breach of legislation'. Choose whichever of these are relevant.

Part 3 – Significant Aspects, Key Issues, and Recommended

Practical Assessment Guidance

Additional Measures

Part 1 – Introduction to the organisation and methodology used

Part 2 – Identifying environmental aspects, associated impacts, and existing controls

Part 3 – Significant aspects, key issues, and recommended additional measures

Part 4 – Communication and review

In the next part of the guidance, we will look at the final part 3 of the NEBOSH form which is concerned with the evaluation of three significant aspects and recommendations to improve them.

The Part 3 Form

Part 3 of the NEBOSH form is shown below:

3. Significant aspects, key issues, and recommended additional measures

You should aim to complete this section in 300 - 400 words for each aspect.

Significant aspect 1:	
Explanation of significance: *Including reference to:* - *environmental receptors that may be affected through associated impacts* - *business concerns, relevant compliance obligations, needs and expectations of interested parties* - *the link to key environmental issues* - *the likelihood and severity of identified impacts (with current controls in place)*	

Recommended additional measures	Intended outcomes	Timescale for implementation	Resource requirement

Name   Learner number

Practical Assessment Guidance

Significant aspect 2:	
Explanation of significance: Including reference to: - environmental receptors that may be affected through associated impacts - business concerns, relevant compliance obligations, needs and expectations of interested parties - the link to key environmental issues - the likelihood and severity of identified impacts (with current controls in place)	

Recommended additional measures	Intended outcomes	Timescale for implementation	Resource requirement

Name   Learner number

Significant aspect 3:	
Explanation of significance: Including reference to: - environmental receptors that may be affected through associated impacts - business concerns, relevant compliance obligations, needs and expectations of interested parties - the link to key environmental issues - the likelihood and severity of identified impacts (with current controls in place)	

Recommended additional measures	Intended outcomes	Timescale for implementation	Resource requirement

Name   Learner number

Note: These forms are for reference purposes only. Please visit the NEBOSH website to obtain the official forms to submit your assessment.

Practical Assessment Guidance

The next task is to choose three significant aspects and evaluate them further. You should take a look at your aspects table and choose the ones with the highest score. Even if you have not classed any of the aspects as having a very high score, it is important that you choose the most significant based on the context of your assessment. You must complete a further more detailed evaluation of the aspects so as to allow a decision to be made on their improvement by a non-specialist. NEBOSH state that for each aspect you must write between 300 and 400 words.

NEBOSH are fairly detailed as to what information you must provide (as you can see on the form). Your significance reasoning must therefore explain:

- What environmental receptors may be affected by the impacts (receptors might include plants, animals, humans, rivers, groundwater etc.).
- Business concerns (such as financial or corporate image), relevant compliance obligations (legal and voluntary requirements) and the needs and expectations of interested parties (stakeholder requirements).
- The key environmental issues with which the impacts are linked - this would include those covered in Unit EMC1, which are:

Key environmental issues
Local effects of pollution (air quality, noise, waste, lighting, odour)
Carbon emissions and the greenhouse effect/global warming
Water resources and ocean pollution
Deforestation, soil erosion and land quality
Material resources and land despoliation, supply chain issues and inequal distribution of impacts
Energy supplies, innovations in food & fuel
Waste disposal and international waste trade
Agricultural issues arising from global trade
Climate change and extreme weather events
Biodiversity loss

Table 2 - from EMC syllabus 1.1

- The likelihood and severity of the impacts, taking into account current control measures employed. You do not need to use a scoring system to outline likelihood and severity - describe using text.

Practical Assessment Guidance

Next the bottom of part 3 of the form for each of the three aspects needs to be completed. This section requires that for each of the three aspects you consider the following:

- **Recommended Additional Measures**

    Before you fill this in, ask yourself if you know what the relevant standards are for each aspect and what all of the control measures would ideally be. So for each aspect: do you know what the legal standard is if there is one? Do you know what any relevant official guidance says? And do you know what best practice looks like?

    You might have a very good understanding of compliance obligations and best practice for a particular aspect. Or you may need to do some research, using your study text and online sources, to discover exactly what the legal standard is for a specific aspect and/or what the best practice guidance is. Again, if you do this kind of research, remember to include it in the methodology section of the Part 1 form.

    Do not write about control measures that are not relevant to the specific aspect in question or are disproportionate to the risks involved. The examiner wants to see that you can identify practical and realistic measures that will address real environmental risks caused by the site's activities. They are not interested in fanciful or over-the-top, risk-averse recommendations, such as shutting down the site for a month as a result of a spillage of oil while secondary containment is fitted.

    Don't just write single words or very short phrases such as 'training' - you will need more information for the description of the measure to be useful. **It is also not recommended to provide general aims or statements such as 'minimise waste'** - think about the actions you will need to take to achieve this aim - these are measures that you should outline in this part of the form.

    Remember that the examiner will be looking at this cell and they will be asking themselves if the additional measures that you have described adequately control the identified aspect.

- **Intended Outcomes**

    You now need to provide a statement as to why you want to implement the recommended additional measures. This must be carried out for each additional measure you propose. The statement is the benefit that the additional measure has, so for example for an additional measure related to reducing energy produced from the combustion of fossil fuel such as 'undertake a review of lighting in the premises', the intended outcome could be 'to reduce energy consumption leading to a reduction in the organisation's contribution to climate change and other emissions'.

- **Timescales for Implementation**

    When you are filling in the 'Timescales for implementation', make sure that you write in lengths of time and not deadline dates. You can use any timescale you like (such as 1 day, 3 days, 1 week, 2 weeks, 3 months, 8 months, 1 year, etc.) but do not put deadline dates (such as 30 September 2022) and do not write 'as soon as possible' or 'ASAP' – this is not a timescale.

    When you are allocating timescales, think about two separate issues:

    - What is the current level of significance presented by the aspect and how urgently is further action required?
        - A poorly controlled aspect presenting a significant impact, where a single action will make a difference to the significance level, will be a high priority.
        - A well-controlled aspect presenting little risk, where an additional control measure has been identified, will be a low priority.

        How easy is it to carry out the additional action from a practical point of view?

    Some actions are very quick and easy to do. They cost little money and can be done within a day. Other actions require major capital expense (which takes time to gain approval for), or they are difficult and time-consuming to do.

    Both of these issues will make a difference to the timescales that you allocate. So an additional measure that is high priority and quick, cheap and easy to do, should be allocated a very short timescale (perhaps 1 or 2 days). Conversely, a lower priority additional measure that is difficult and time-consuming should be allocated a longer timescale (perhaps months).

Practical Assessment Guidance

The key issue is that the examiner should be able to see that the timescales you are setting look practical and realistic and, most importantly, address the significance presented by each aspect in a proportionate way.

- **Resource Requirement**

    In this section you need to roughly estimate the financial cost of implementing the recommendation. For some tasks that do not have a direct financial cost, you could estimate the amount of time that it might take to complete.

Part 4 – Communication and Review

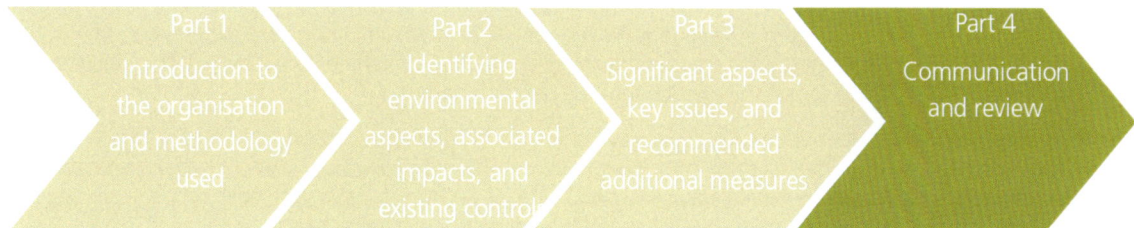

In the final part of the Practical Assessment Guidance, we will look at Part 4 of the assessment which requires you to indicate when you intend to review your aspect assessment, and how you intend to communicate the findings and follow up the assessment to check that actions have been carried out.

The Part 4 Form

Part 4 of the NEBOSH form is shown below:

Communication and review

You should aim to complete this section in 50 - 100 words.

Timescale for review	
Who should review these recommendations and how you will communicate to them	
Follow-up procedure(s)	

Name   Learner number

Note: These forms are for reference purposes only. Please visit the NEBOSH website to obtain the official forms to submit your assessment.

Practical Assessment Guidance

The last step in completing the practical assessment is to complete Part 4 of the assessment form. NEBOSH state that you should aim to complete this section in 50 to 100 words.

This part of the form is split up into three rows:

- In the first row, you must give an aspect and impact assessment review date and explain why you have chosen that date. It is probably sensible to state a timescale for the review (e.g. 1 year) and state the date that this would be on relative to the date that you put at the top of the assessment form (e.g. 31 August 2021). You must justify your choice of review date so that the examiner can understand why you picked that date.

- In the second row, you must explain who you will need to review the findings and how you will communicate the findings of the assessment. A range of people may need to know the result of the assessment (for example, the workers directly involved, responsible managers, directors and those with specific actions allocated to them). The results might be communicated in various ways (by issuing hard copies of the assessment, through toolbox talks or safety briefings, in training sessions, via one-to-one meetings, via the organisation's intranet system, etc.). You must only state the role in this section, not the person's name.

- In the third and last row, you must explain how you intend to follow up your aspect assessment to check that all of the identified actions have been carried out. You should think about how you will track each action forward, how you will set reminders before each action becomes due, how you will liaise with each responsible person to keep track of overdue actions and how you will escalate long overdue actions that do not appear to have been adequately addressed.

The key issue is that the examiner must see that, having completed the aspect assessment, you are able to propose sensible and realistic follow-up action that will yield practical results.

Before you finish with Part 4 and move on to the final submission of your practical assessment, check to ensure that you have addressed all of the information requested on Part 4 of the form.

Final Submission

Don't forget to check that your name and learner number are written on all parts of the form that you submit.

Once you have filled in all parts of the form, you have completed the practical assessment, and can then submit it for marking. You will receive submission instructions in due course via email. NEBOSH will mark the assessment and send your results directly to you.

Please note that NEBOSH reserves the right to submit your assessment to a plagiarism detection software package.

Cases of suspected plagiarism will be investigated by NEBOSH and proven cases will be dealt with in line with their malpractice policies.

A worked example of a completed practical assessment is presented below. Additional useful information and some summary tips are included in the Final Reminders section at the end of this Unit.

Practical Assessment Guidance

Worked Example

NEBOSH have included a worked example of the Unit EMC2 practical assessment in their Assessment Pack so that you can see what you are aiming for. It is duplicated as a worked example here.

1. Introduction to the organisation and methodology used

You should aim to complete this section in 200 - 250 words.

Name of organisation*	NPD University
Site location (town / area)	Neboshville
Number of workers	450 staff across the whole university, a mixture of full/part time faculty as well as non-teaching staff.
General description of the organisation	NPD is a small university with the campus located on the outskirts of the city. The university campus facilities are also used by the public for sports events, as well as other regular talks / events in lecture halls and the library.
Description of the area or process to be included in the assessment	There are five café booths across campus serving takeaway hot drinks (teas and coffees), cold drinks in bottles/cans, and packaged snack foods. All booths are owned by the university and manned by directly employed university workers. These are located on the ground floor of University buildings close to offices, classrooms, and lecture theatres. One cafe is located in the library building. Most of the café's customers work or study at the university, although some locations are also open to the public. Opening times vary by location and some are closed over the weekend. Each booth has one or two large open refrigerators which contain cold drinks and food items. Each has a large coffee machine and hot water dispenser. Behind the counter there is a small storage area, which leads to shared service / waste collection areas behind the buildings. Each booth is located next to an open seating area, which are available for anyone to use.
Any other relevant information	The University has a facilities management team who have day-to-day responsibility for environmental issues. The catering manager oversees the café booths as well as all other catering facilities in the university campus. There is a full-time manager of each booth along with several part-time workers – usually students at the university. Legal compliance is assessed by the facilities management team; the main environmental issues which catering activities contribute to are waste and use of refrigerants. Environmental performance criteria also now form part of the national 'University Guide' which is highly publicised and can positively or negatively affect the number of new student applications. Senior management are therefore keen to make and publicise improvements wherever possible.

* *If you're concerned about confidentiality, you can invent a false name and location for your organisation. All other information provided must be factual.*

Ann Learner   0000001

Practical Assessment Guidance

You should aim to complete this section in 100 - 200 words.

| How the aspect and impact assessment was carried out, including:
- sources of information consulted;
- who you spoke to; and
- how existing controls were identified | I spoke with the catering manager to get a general overview of the café booths, locations, volume of sales etc. I then visited a couple of the locations during a busy morning period to observe activities that took place. Whilst there I spoke to one of the café managers and one of the part-time workers, for more information on the day-to-day running.

After my initial observations I looked at a few internal documents for more information on the controls which are currently in place:

- NPD Environmental policy (this applies across the whole university and is available on the public website);
- waste management procedures, equipment inspection records, and monitoring records (requested from facilities management team); and
- procurement policy / supplier approval process (received from catering manager).

I also looked at last year's national 'University Guide' to see what environmental criteria were considered, and for beneficial practices done by other Universities that we might be able to implement.
I did a small amount of general online research when making my recommendations. For example, I looked at some environmental labelling schemes that are used on tea and coffee, and possible alternative packaging options. |

Note: this section can be completed after you have competed your full assessment.

2. Part 2 - Identifying environmental aspects, associated impacts, and existing controls

Activity, product, or service	Aspects	Operating condition(s)	Associated impacts	Existing controls	Significance rating	Justification for each significance rating
Packaged food / cold drinks						
Refrigerators – running	electricity usage	Normal	Depletion of non-renewable resources for electricity generation	none	**3 (significant)**	'aspect is not fully controlled under normal conditions' 'multiple instances of same issue'
	refrigerant usage	Normal	Ozone depletion / global warming in event of leak	Register of refrigerants has been compiled and is regularly reviewed to ensure only legal refrigerants are used Contracted maintenance company carry out servicing and refrigerant leak testing annually	**3 (significant)**	'potential breach of legislation' 'multiple instances of same issue'
Refrigerators - disposal	Hazardous (electrical) waste	Abnormal	Ozone depletion / global warming in event of leak Land contamination in event of waste to landfill	Waste is stored securely before collection Suitable waste contractors approved/used for disposal	**3 (significant)**	'potential breach of legislation' 'business threat from impacts' (if not disposed of properly)
Food / drink sourcing	Resource use / environmental damage during manufacture	Normal	Depletion of natural resources Deforestation Changes in land use leading to biodiversity loss	Purchasing policy is in place for entire business to favour products with environmental certification and/or that are sourced from local businesses	2 (minor)	'aspect controlled under normal conditions) 'no potential for breach of legislation or other compliance obligation'
	Fuel use as part of distribution/delivery network	normal	Generation of greenhouse gases / contribution to global warming			
Food / drink packaging	Generating waste on and offsite - Delivery packaging - Consumer waste - Litter on campus	Normal	Depletion of resources Waste to landfill / incineration Local water / land pollution	Waste is separated into 'recyclable' and 'general waste' Waste is stored securely before collection	**3 (significant)**	'aspect is not fully controlled under normal conditions'

Ann Learner 0000001

Practical Assessment Guidance

Activity, product, or service	Aspects	Operating condition(s)	Associated impacts	Existing controls	Significance rating	Justification for each significance rating
				Licenced waste carriers used for all types of waste		'multiple instances of the same issue / large volumes involved'
Hot drinks						
Single-use cups	Use of natural resources during manufacture and distribution	Normal	depletion of natural resources	None - offered as standard to all customers (customers own reusable cups accepted but not promoted / encouraged)	3 (significant)	'Aspect is not currently controlled under normal conditions' 'interested parties aware of issue'
	Generating waste on and offsite - Delivery packaging - Consumer waste (non-recyclable)	normal	Waste to landfill / incineration Local water / land pollution	'Recyclable' and 'general waste' bins available Waste is stored securely before collection Licenced waste carriers used for all types of waste	3 (significant)	'large volumes involved'
Single-use / individual items available to customers (stirrers, sachets) at counter	Use of natural resources during manufacture and distribution	Normal	depletion of natural resources	Purchasing policy is in place for entire business to favour products with environmental certification	3 (significant)	'aspect is not fully controlled under normal conditions' 'interested parties aware of issue'
	Generating waste on and offsite - Delivery packaging - Consumer waste - Litter on campus	normal	Waste to landfill / incineration Local water / land pollution	'General waste' bins available at counters Waste is stored securely before collection Licenced waste carriers used for all types of waste	3 (significant)	'large volumes involved'

Ann Learner 0000001 5

Activity, product, or service	Aspects	Operating condition(s)	Associated impacts	Existing controls	Significance rating	Justification for each significance rating
Consumables – sourcing (coffee, teabags, milk)	Resource use / environmental damage during manufacture Fuel use as part of distribution/delivery network	Normal	Depletion of natural resources Deforestation Changes in land use leading to biodiversity loss Global warming	Purchasing policy is in place for entire business to favour products with environmental certification and/or that are sourced from local businesses	2 (minor)	'aspect controlled under normal conditions) 'no potential for breach of legislation or other compliance obligation'
Consumables – disposal (coffee, teabags)	Large volumes of food waste Generation of greenhouse gases (during decomposition)	Normal	Waste to landfill Global warming	'Recyclable' and 'general waste' bins available No guidance on food waste Waste is stored securely before collection Licenced waste carriers used for all types of waste	3 (significant)	'aspect is not fully controlled under normal conditions' 'large volumes involved'
Promotional displays / branding						
Replacing seasonal signage	Paper use Ink use Electricity use Generation of greenhouse gases as part of the distribution/delivery network	Normal	Depletion of natural resources for energy use Deforestation Global warming	None	3 (significant)	'aspect is not fully controlled under normal conditions'
	Waste generation (Mixture of recyclable and non-recyclable printed materials)	normal	Waste to landfill	None - no guidance on what materials are recyclable		
Branded uniforms – new for all team members	Raw materials for textile manufacture Dye / processing Electricity use during embroidery (branding)	Abnormal	Depletion of natural resources Deforestation Changes in land use leading to biodiversity loss Global warming	Purchasing policy is in place for entire business to favour products with environmental certification	2 (minor)	'no potential for breach of legislation or other compliance obligation'

Ann Learner 0000001 6

© RRC International — Unit EMC2 - Practical Assessment Guidance — 17

Practical Assessment Guidance

Activity, product, or service	Aspects	Operating condition(s)	Associated impacts	Existing controls	Significance rating	Justification for each significance rating
	Fuel use as part of distribution/delivery network					
	Textile waste generated	Abnormal	Waste to landfill	Currently no guidance for staff on disposal		'no potential for breach of legislation or other compliance obligation' 'small quantities involved'

3. **Significant aspects, key issues, and recommended additional measures**

You should aim to complete this section in 300 - 400 words for each aspect.

Significant aspect 1: Electricity usage (refrigerators – running)

Explanation of significance: Including reference to: - environmental receptors that may be affected through associated impacts - business concerns, relevant compliance obligations, needs and expectations of interested parties - the link to key environmental issues - the likelihood and severity of identified impacts (with current controls in place)	This aspect of fridges running is present under normal operating conditions and is a key part of the function of the café booths, as most of the cold foods, snacks and drinks are stored there. Electricity usage and it's impacts is an issue as the equipment is present in all locations covered in this assessment, and this usage contributes to the university's wider statistics. The electricity used by the university is produced from non-renewable carbon, which is a depleting natural resource, and also produces CO_2 - a greenhouse gas which contributes to global warming. Global warming is a current focus internationally as an urgent environmental issue, with national targets to minimise temperature increases. The severity of global warming has been demonstrated, with its knock-on effects to many other global environmental issues including damage to ecosystems as well as rising sea levels. The refrigerators run all day and night. Three of the five café booths are closed over the weekend; however the fridges are left running. Café staff indicated that they have never been instructed to turn them off over the weekend, but there would be no issue with this (based on what items are left over the weekend). This would immediately reduce the energy consumption for the cafes. The refrigerators are open cabinets which means that cold air is constantly escaping. These particular models are relatively old and therefore are likely to be replaced soon; a different style of fridge with doors is recommended to reduce energy use further. There is the potential to reduce impacts still by considering energy from renewable sources, or even on-site generation (this is a wider issue for the organisation as a whole). This would reduce reliance on non-renewable resources, and likely reduce CO_2 emissions.

Recommended additional measures	Intended outcomes	Timescale for implementation	Resource requirement
Change procedure for locations which close over the weekend to turn off refrigerators.	Reduce energy usage, and associated emissions.	Immediate	Zero cost to change procedures
Consider increased monitoring in order to determine benefit/reduction in usage	Be able to quantify benefits from the changes made	1-2 months	Time – facilities team (if monitoring is increased)
Select replacements with glass doors to reduce energy and refrigerant consumption	Greatly reduce usage at all five locations	Replacement fridges – as required, based on current equipment will likely be replaced within 12 months	Financial cost of replacing fridges at all five locations = $5000 (rough estimate). Cost could be spread through gradual replacement

Ann Learner   0000001   8

Practical Assessment Guidance

Significant aspect 2: Waste generation (single-use / individual items available to customers (stirrers, sachets))	
Explanation of significance: Including reference to: - environmental receptors that may be affected through associated impacts - business concerns, relevant compliance obligations, needs and expectations of interested parties - the link to key environmental issues - the likelihood and severity of identified impacts (with current controls in place)	The hot drinks available from the café booths are considered 'takeaway only'. Individual sachets (sugar/sweeteners), and disposable stirrers are therefore available for customers to take as required. The average customer takes one stirrer and three sachets per hot drink, across all five locations this adds up to incredibly large volumes within normal operation. The main issue associated is the generation of waste. Although the wooden stirrers are biodegradable, they are put into 'general waste' as they are contaminated with food and as such cannot be recycled. The sachet packets are recyclable however there are no segregated bins at the point of use and so everything goes into the general waste, which ultimately goes to landfill – likelihood of waste impact is therefore high. Waste-to-landfill is also monitored and reported on by the facilities team, recycling volumes and targets are also reported on by the university as a whole. Small items like this also contribute to litter around campus, which has a more direct environmental impact. Litter on land often ultimately ends up entering the watercourse and resulting in pollution, posing a threat to plant and animal life. This impact is also cumulative, and although relatively small amounts of litter arise from this activity, it is contributing to wider problems which has serious and far-reaching consequences. There is increasing awareness among the student body around single-use and disposable items. The booth manager I spoke to informed me that customers regularly mention this (a recent change). I observed that the vast majority of customers used these at the counter before leaving with their drinks. The recommendation is therefore to replace the single use items – provide spoons instead of stirrers, and replace sachets with 'pourers' – these are already used in other catering locations at the university. This could be trialled, maintaining a reduced supply of sachets to be available on request only. The catering department has a purchasing policy which already favours products with an environmental certification; this usually refers to the food/agricultural practices. It is recommended that the policy is expanded to include packaging materials, to help with sourcing in the future, and ensure that improvements are still made, even if it is determined that some single-use items are still required.

Recommended additional measures	Intended outcomes	Timescale for implementation	Resource requirement
Replace sachets with re-fillable 'pourers', and replace wooden stirrers with (washable) spoons (already used at other catering functions)	Minimise requirement for single-use item, reducing recyclable and non-recyclable waste	1 month	Financial – relatively low cost to purchase re-usable items. Cost is offset by better value of bulk-buying sugar etc (already purchased for other catering functions) Cost of bins – likely zero as these can be relocated from
Where replacement is not suitable, source alternatives from sustainable sources and which have recyclable packaging	Where waste cannot be eliminated, allows it to be re-cycled. Contributes to university waste targets.	1-2 months 1 month	

Provide split bins at counter with signage to show what packaging can be recycled Update purchasing policy to assist in sourcing decisions	Ensures proper segregation of waste; further contributing to waste targets / reduced waste to landfill Increase awareness and understanding of environmental considerations; provide information for decision-makers	1-2 months	elsewhere. Signage can be produced / printed internally Time – catering manager to research suitable alternatives and update policy

Practical Assessment Guidance

Significant aspect 3: Food waste (consumables – disposal (coffee, teabags))	
Explanation of significance: *Including reference to:* - environmental receptors that may be affected through associated impacts - business concerns, relevant compliance obligations, needs and expectations of interested parties - the link to key environmental issues - the likelihood and severity of identified impacts (with current controls in place)	The sale of hot drinks produces large volumes of food waste (tea and coffee). Currently this is all disposed of as general waste. Conversations with the catering manager and facilities team identified that there is an opportunity for this to be added to the on-site composter which is used on campus. Landfill as ultimate waste disposal is a key environmental issue as there is a finite amount of space available for sites to operate. Local disruptions are also an issue when sites are in use, as well as limited possibilities for land use once the landfill site is 'full'. Food waste being sent to landfill produces methane, which is a harmful greenhouse gas – contributing to global warming. The likelihood of this impact resulting from these activities is high, as there is currently no alternative option available for disposal. The severity of landfill impacts increases over time, as locations become harder to find and manage. Composting food waste is an earlier stage of the waste hierarchy, which not only reduces methane production from the same waste but will also reduce costs to the organisation (for waste removal, and by producing compost on-site). When I raised this suggestion to the booth managers, they implied that teabags could not be composted. After a quick check on the packaging it was confirmed that the current brand used is actually labelled as compostable. The changes suggested are relatively minor in practice; however, it will have a large impact on the organisation's environmental performance, based on the large volumes of waste involved.

Recommended additional measures	Intended outcomes	Timescale for implementation	Resource requirement
Separate compostable waste for collection by facilities team (procedure change)	Ensures proper segregation of waste; further contributing to waste targets; reduced waste to landfill; reduced costs of disposal	1-2 weeks	Cost of bins – likely zero as these can be relocated from elsewhere. Signage can be produced / printed internally.
Communication to workers to clarify which items are compostable / non-recyclable	Increase awareness and understanding of environmental considerations, enable workers to better contribute to environmental objectives	1 week	Zero cost to change procedures
Update purchasing policy to assist in sourcing decisions	Increase awareness and understanding of environmental considerations; provide information for decision-makers	1-2 months	Time – catering manager to research suitable alternatives and update policy

Ann Learner 0000001

Communication and review

You should aim to complete this section in 50 - 100 words.

Timescale for review	Recommend annual review of the assessment as working procedures are unlikely to change / no planned changes.
Who should review these recommendations and how you will communicate to them	Set up brief meeting with facilities manager and catering manager to advise on findings and recommendations. They can then distribute information and actions in their departments.
Follow-up procedure(s)	Follow up in 1-2 months to check on implementation/effectiveness of recommended short-term actions, and allow for any queries on continuing actions.

Ann Learner 0000001

Note: This sample is for reference purposes only. Please visit the NEBOSH website to obtain the current version of the sample.

Please do not copy any part of this worked example as this would constitute plagiarism that would be investigated by NEBOSH and dealt with in line with their malpractice policies.

Practical Assessment Guidance

Final Reminders

This section contains some reminders to help you with the practical assessment. It offers advice on how to approach the assessment and has some useful hints and tips.

Summary Guidance on the Practical Assessment

The practical assessment requires you to carry out an aspect and impact assessment at your workplace. There are several parts to this assessment as shown in the diagram below:

The images shown are for illustrative purposes only and are subject to change by NEBOSH.

You must record your assessment on the standard form supplied by NEBOSH in its online Assessment Pack. You can download this form from their website. Forms should be completed electronically.

Make sure that you include all of the information that is requested at the top of each and every form.

Part 1 - Introduction to the Organisation and Methodology Used

Practical Assessment Guidance

1. **Introduction to the organisation and methodology used**

You should aim to complete this section in 200 - 250 words.

Name of organisation*	
Site location (town / area)	
Number of workers	
General description of the organisation	
Description of the area or process to be included in the assessment	
Any other relevant information	

* If you're concerned about confidentiality, you can invent a false name and location for your organisation. All other information provided must be factual.

Name _____ Learner number _____

Description of the Organisation

Ordinarily the workplace that you choose to describe would be the organisation that you work for. It needs to be large and complex enough to present a good range of aspects.

You can provide a false name and/or location for your organisation if you like, to protect the identity of the organisation and maintain confidentiality.

You must include the organisation's name and location (even if they are fictitious). Include the number of workers and the shift patterns that are worked. Give a good general description of what the organisation does and the site layout so that the examiner can picture the kind of workplace, the products or services involved, and the sorts of activities that are likely to be carried out there in addition to vulnerable receptors.

Methodology Used

You should include things such as the sources of information that you consulted, who you spoke to, and how the aspects and additional measures were identified.

Practical Assessment Guidance

Part 2 - Identifying Environmental Aspects, Associated Impacts, and Existing Controls

Part 1 Introduction to the organisation and methodology used | **Part 2** Identifying environmental aspects, associated impacts, and existing controls | **Part 3** Significant aspects, key issues, and recommended additional measures | **Part 4** Communication and review

2. Part 2 - Identifying environmental aspects, associated impacts, and existing controls

Activity, product, or service	Aspect	Operating condition(s)	Associated impacts	Existing controls	Significance rating	Justification for each Significance rating

Please note that Part 2 of the aspects and impacts assessment form does not give any indication of word count.

You must include at least **four** different **activities/products/services** from your chosen site. The activity, product or service selected should be large enough for meaningful examination and small enough to be sufficiently understood.

You need to identify at least **10 different environmental aspects** in your assessment with at least one associated impact (often there will be more than one impact and you should include all of them).

Issues that are commonly addressed include:

- Emissions to air.
- Releases to water.
- Waste management.
- Contamination of land.
- Use of raw materials and natural resources.

Practical Assessment Guidance

Next you need to consider the operating conditions of each aspect. T**o pass the assessment you must include two of the following conditions**:

- Normal.
- Abnormal.
- Accidents/incidents/emergencies.

The next stage is to determine environmental impacts of each aspect. These are the changes to the environment that are caused by the aspect.

The next part of the form requires you to provide information on the current controls for each impact(s). You need to write a brief explanation of all of the current control measures that are in place. Don't just write single words or very short phrases such as 'training' and don't simply refer to one control repeatedly as the only control measure.

Now you have identified the aspects and impacts, you need to determine which are significant. NEBOSH provides a significance rating table to assist in making this decision. **You must use this table to evaluate your aspects and impacts**:

Practical Assessment Guidance

Justification for significance rating	Significance rating
<ul><li>Aspect is not currently controlled under normal condition</li><li>'Abnormal' condition would breach regulations</li><li>Evidence of / high likelihood of immediate environmental damage</li><li>Severity of impacts is extensive</li><li>In breach of legislation</li><li>Interested parties have voiced concerns</li><li>Repeated complaints received (internal/external)</li></ul>	4 (Major) Immediate action and control required
<ul><li>Aspect is not fully controlled under normal conditions</li><li>Severity of 'normal' condition impacts are moderate</li><li>Abnormal condition impacts are likely to occur and/or unlikely to be detected</li><li>Impacts are likely to increase based on planned activities</li><li>Emergency situation would cause major environmental impact</li><li>Potential breach of legislation</li><li>In breach of company policy or other non-legal compliance obligation</li><li>Financial / business threat from impacts</li><li>Interested parties aware of issue</li><li>Complaint received (internal/external)</li><li>Multiple instances of same issue / large volumes involved</li></ul>	3 (Significant) Action required within set timeframe, with scheduled review of effectiveness
<ul><li>Aspect controlled under normal conditions</li><li>Abnormal condition impacts are unlikely to occur and/or likely to be detected</li><li>Severity and likelihood of any impact is low</li><li>No potential for breach of legislation or other compliance obligation</li><li>Single instance of issue / small quantities involved</li></ul>	2 (Minor) Additional actions possible; recommend to review in the future
<ul><li>Aspect controlled under normal and abnormal conditions</li><li>No additional impact arising from emergency condition</li><li>Severity and likelihood of any impact is minimal</li><li>Very small quantities involved</li></ul>	1 (Negligible) Limited additional action possible; review in the future may be required

Note: meeting one of the criteria is enough for the aspect to be considered at the highest rating.

For each aspect you need to choose the significance rating that best reflects the aspect. Your aspect only needs to meet one criteria in a category for it to be considered at the highest rating. Place in the 'Significance rating' column on the form the score and description of the significance rating, for example 4 (Major) or 2 (Minor).

In the criteria column you need to copy across from the significance rating table the reason why you have chosen the rating from the bullet-pointed criteria list, so this might be for 3 (Significant) 'Potential breach of legislation'.

Practical Assessment Guidance

Part 3 – Significant Aspects, Key Issues, and Recommended Additional Measures

Part 1 Introduction to the organisation and methodology used

Part 2 Identifying environmental aspects, associated impacts, and existing controls

Part 3 Significant aspects, key issues, and recommended additional measures

Part 4 Communication and review

Part 3 of the NEBOSH form is shown below:

3. **Significant aspects, key issues, and recommended additional measures**

You should aim to complete this section in 300 - 400 words for each aspect.

Significant aspect 1:	
Explanation of significance: Including reference to: - environmental receptors that may be affected through associated impacts - business concerns, relevant compliance obligations, needs and expectations of interested parties - the link to key environmental issues - the likelihood and severity of identified impacts (with current controls in place)	

Recommended additional measures	Intended outcomes	Timescale for implementation	Resource requirement

Name　　　　　Learner number

Practical Assessment Guidance

Significant aspect 2:	
Explanation of significance: Including reference to: - environmental receptors that may be affected through associated impacts - business concerns, relevant compliance obligations, needs and expectations of interested parties - the link to key environmental issues - the likelihood and severity of identified impacts (with current controls in place)	

Recommended additional measures	Intended outcomes	Timescale for implementation	Resource requirement

Name: Learner number:

Significant aspect 3:	
Explanation of significance: Including reference to: - environmental receptors that may be affected through associated impacts - business concerns, relevant compliance obligations, needs and expectations of interested parties - the link to key environmental issues - the likelihood and severity of identified impacts (with current controls in place)	

Recommended additional measures	Intended outcomes	Timescale for implementation	Resource requirement

Name: Learner number:

Practical Assessment Guidance

The next task is to choose three significant aspects and evaluate them further.

You should take a look at your aspects table and choose the ones with the highest score. Even if you have not classed any of the aspects as having a very high score, it is important that you choose the most significant based on the context of your assessment. You must complete a further more detailed evaluation of the aspects so as to allow a decision to be made on their improvement by a non-specialist.

NEBOSH are fairly detailed as to what information you must provide. Your significance reasoning must therefore explain:

- What environmental receptors may be affected by the impacts.
- Business concerns.
- The key environmental issues with which the impacts are linked such as:

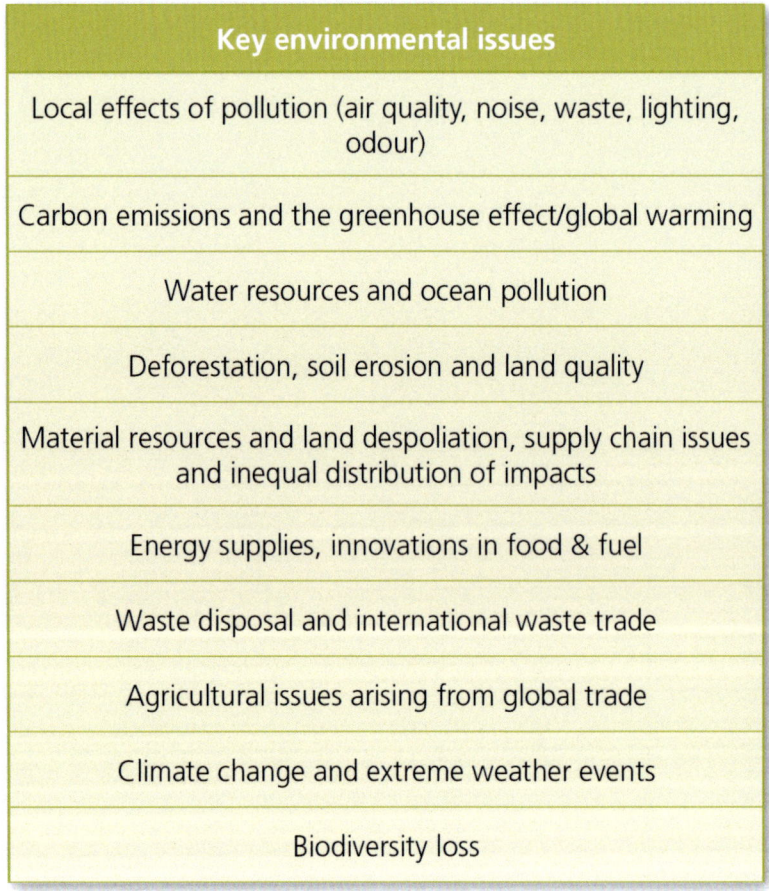

Key environmental issues
Local effects of pollution (air quality, noise, waste, lighting, odour)
Carbon emissions and the greenhouse effect/global warming
Water resources and ocean pollution
Deforestation, soil erosion and land quality
Material resources and land despoliation, supply chain issues and inequal distribution of impacts
Energy supplies, innovations in food & fuel
Waste disposal and international waste trade
Agricultural issues arising from global trade
Climate change and extreme weather events
Biodiversity loss

Table 2 - from EMC syllabus 1.1

- The likelihood and severity of the impacts, taking into account current control measures employed. You do not need to use a scoring system to outline likelihood and severity - describe using text.

Practical Assessment Guidance

Next the bottom of part 3 of the form for each of the three aspects needs to be completed. This section requires that for each of the three aspects you consider the:

- **Recommended Additional Measures**

    Don't just write single words or very short phrases such as 'training' - you will need more information for the description of the measure to be useful. **It is also not recommended to provide general aims or statements such as 'minimise waste'** - think about the actions you will need to take to achieve this aim - these are measures that you should outline in this part of the form.

- **Intended Outcomes**

    You now need to provide a statement as to why you want to implement the recommended additional measures.

- **Timescales for Implementation**

    When you are filling in the 'Timescales for implementation', make sure that you write in lengths of time and not deadline dates. You can use any timescale you like (such as 1 day, 3 days, 1 week, 2 weeks, 3 months, 8 months, 1 year, etc.) but do not put deadline dates (such as 30 September 2022) and do not write 'as soon as possible' or 'ASAP' – this is not a timescale.

- **Resource Requirement**

    In this section you need to roughly estimate the financial cost of implementing the recommendation. For some tasks that do not have a direct financial cost you could estimate the amount of time that it might take to complete.

Part 4 – Communication and Review

Part 1	Part 2	Part 3	Part 4
Introduction to the organisation and methodology used	Identifying environmental aspects, associated impacts, and existing controls	Significant aspects, key issues, and recommended additional measures	Communication and review

Communication and review

You should aim to complete this section in 50 - 100 words.

Timescale for review	
Who should review these recommendations and how you will communicate to them	
Follow-up procedure(s)	

Note: These forms are for reference purposes only. Please visit the NEBOSH website to obtain the official forms to submit your assessment.

Practical Assessment Guidance

This part of the form is split up into three rows:

- In the first row, you must give an aspect and impact assessment review date and explain why you have chosen that date.
- In the second row, you must explain who you will need to review the findings and how you will communicate the findings of the assessment. You must only state the role in this section, not the person's name.
- In the third and last row, you must explain how you intend to follow up your aspect assessment to check that all of the identified actions have been carried out.

The key issue is that the examiner must see that, having completed the aspect assessment, you are able to propose sensible and realistic follow-up action that will yield practical results.

HINTS AND TIPS

Do:

- Read all parts of the official NEBOSH guidance on the practical assessment (available from the NEBOSH website) and the guidelines in this study text before you submit your forms.
- Complete the assessment gradually rather than in a rush just before the deadline for submission.
- Read all of the relevant content of the course before preparing your assessment forms so that you have a good understanding of the nature of aspects and impacts and ways to control them.
- Carefully check your work for spelling errors and simple technical errors by double checking in the study text and online.
- Make sure that you put the right amount of work into the relevant parts of the assessment. Use the word count given by NEBOSH as an indication of how much work to do.
- Check that you have written your name and learner number on every part of every form where required.
- Check that your final document for submission contains all of the relevant parts and that you have not accidentally missed a part out.
- Look at the Worked Example published by NEBOSH which is reproduced above and is also available from the NEBOSH website.

Practical Assessment Guidance

HINTS AND TIPS

Do not:

- Carry out your assessment in a very restricted workplace with too few aspects.
- Ignore the information requested at the top of every page of the form.
- Write a very short, incomplete introduction to your workplace or explanation of your methodology.
- Pick fewer than four different activities/products/services from your chosen area.
- Pick fewer than 10 different environmental aspects for your aspect assessment.
- Pick only one of the following conditions: normal, abnormal, accidents/incidents/emergencies.
- Choose one impact for an aspect when there is more than one.
- Use another type of significance rating other than the one provided by NEBOSH.
- Repeat the aspect and impacts over and over.
- Give a very brief, unspecific description of the current measures in place or the additional control measures required (such as by using single words such as 'training').
- Give 'ASAP' as the timescale for your actions.
- Pick low risk, trivial or well-controlled aspects for your three top priority significant aspects.
- Write very short and unconvincing justifications for why you picked the top three aspects that you did.
- Fail to identify a review date.
- Leave the organisation's name, aspect assessment date or assessment scope off the Part 1 aspect assessment form.
- Leave your learner number and learner name off any parts of the form.
- Copy any parts of your assessment from any other student or from any other source. You do not want to get investigated for plagiarism and fall foul of NEBOSH's malpractice rules.
- Copy any parts of the Worked Example (mentioned above). NEBOSH will be alerted to this and will treat it as plagiarism.

Good luck with the practical assessment!

Unit EMC1

Suggested Answers

No Peeking!

Once you have worked your way through the Study Questions in this book, use the Suggested Answers on the following pages to find out what you got right (and where you went wrong) to check your understanding.

Suggested Answers to Study Questions

Element 1: Foundations in Environmental Management

Question 1
The three media that make up the 'environment' are air, water and land.

Question 2
Much of the solar radiation that strikes the planet surface is reflected back towards space. Although carbon dioxide, water vapour, methane, chlorofluorocarbons (CFCs) and some other gases in the atmosphere are transparent to visible light, they intercept and absorb much of the reflected infrared radiation, re-reflecting it back towards Earth. This process retains some of the solar heat and is called the 'greenhouse effect'.

When greenhouse gases build up in the atmosphere, more heat is trapped near the Earth's surface. Ocean surface temperatures rise, so more water vapour enters the atmosphere and the Earth's surface temperature rises more. This is 'global warming'.

Question 3
The three main reasons why organisations need to manage environmental impacts are:

- Ethical (or moral).
- Legal.
- Financial (or economic).

Question 4
Legal and economic effects that could occur following a pollution incident include:

- Cost of fines.
- Clean-up costs.
- Compensation payments.
- Indirect costs from loss of credibility and support in the market.

(Only three were required.)

Question 5
Achieving sustainable development requires four objectives to be met:

- Effective protection of the environment.
- Prudent use of natural resources.
- Maintenance of stable levels of growth.
- Social progress.

These are often conflicting objectives. For example:

- 'Prudent use of natural resources' could involve no longer extracting fossil fuels. This would result in the loss of the most commonly used energy source in the developed world. Without dramatic technological advances, we would be unable to drive vehicles and manufacture products. In such an extreme example, the economy is likely to collapse and there would be social regression rather than progression.
- 'Social progress' of developing countries is likely to lead to more damage to the environment, but social progress is an objective of sustainable development.

Question 6
(a) The Montreal Protocol on Substances that Deplete the Ozone Layer.

(b) The Basel Convention on the Control of Transboundary Movements of Hazardous Wastes and their Disposal.

Suggested Answers to Study Questions

Question 7

BPEO is defined as "the outcome of a systematic consultative and decision-making procedure which emphasises the protection and conservation of the environment across land, air and water. The BPEO procedure establishes for a given set of objectives, the option that provides the most benefits or the least damage to the environment, as a whole, at acceptable cost, in the long term as well as in the short term".

Question 8

Environmental permits are often required for:

- Discharge to groundwater.
- Keeping, treating and disposing of waste.
- Emissions of pollutants to air.
- Operations where more than one activity includes a permit (when an integrated permit is used).

Suggested Answers to Study Questions

Element 2: Environmental Management Systems

Question 1

Implementation of an EMS can involve substantial resources, such as time, money, facilities and people. Such resources are only likely to be available if management are committed to the project and allocate the resources.

If management are known and seen to be committed to the project, other people are more likely to take time out from their other priorities and objectives to provide information and assistance critical to the implementation of the EMS.

If management support the project as being a priority, all staff will treat it as a priority.

Management are key to the dissemination of information and to training and educating the workforce. If they are committed to the project, they will help all in the organisation to understand the concept and importance of environmental management.

Question 2

An environmental policy is a public declaration by the senior management of an organisation of their commitment to protecting the environment. It should set out the organisation's intentions regarding the environment and is the foundation upon which an Environmental Management System (EMS) can be built.

It must reflect the reality of the activities that the organisation undertakes and it must be a catalyst for action by the organisation. An effective and active policy provides an environmental framework within which all decisions can be made.

Question 3

The seven key stages within an ISO 14001 environmental management system are:

- Context of the organisation.
- Leadership.
- Planning.
- Support.
- Operation.
- Performance evaluation.
- Improvement.

Question 4

Responsibility for preparing and endorsing a written environmental policy rests with the organisation's senior director or senior manager, who is also responsible for ensuring that it is implemented, reviewed at appropriate intervals and updated when necessary.

Suggested Answers to Study Questions

Question 5

The following are the main inputs to the management review:

- Results of internal audits and evaluations of compliance with legal and other requirements.
- Communication from interested external parties, including complaints.
- The environmental performance of the organisation.
- The extent to which objectives and targets have been met.
- The status of preventive and corrective actions.
- Follow-up actions from previous management reviews.
- Changing circumstances, including developments in legal and other requirements.
- Recommendations for improvement.

(Only three were required.)

Question 6

The frequency of the inspections will be dependent on issues such as:

- The purpose of the inspection.
- Any frequency imposed by regulations, such as discharge consents and environmental permits.
- The level of risk to the environment.
- Conditions found at the last inspection.

Question 7

The minimum requirements for someone carrying out an environmental inspection are as follows:

- An understanding of the tools of workplace inspections, their advantages and disadvantages and how to use them.
- An understanding of the process or activity being inspected.
- Knowledge of the potential environmental impacts from the process or activity.
- Knowledge of the standards that are acceptable.
- A basic report-writing ability or ability to use a checklist.

Question 8

Audits are a pre-planned, systematic and objective assessment of a situation against a given set of criteria. For example, consider the following as a criterion: Clause 5.2 of ISO 14001:2015 states that "top management shall establish, implement and maintain an environmental policy that, within the defined scope of its environmental management system ... is appropriate to the purpose and context of the organisation, including the nature, scale and environmental impacts of its activities, products and services". Thus, against this particular point, the auditor would be looking for evidence to establish that a policy is in place, has been agreed by top management and is appropriate to the nature and scale of the organisation.

Inspections are an assessment of what is there at the time an inspection takes place. They are essentially looking for signs of failure, e.g. leaks in pipes, broken lights, signs of damage, etc.

Suggested Answers to Study Questions

Question 9

You could have chosen any of the following benefits and limitations associated with the implementation of an EMS and certification to the ISO 14001 standard:

Benefits

- Increased compliance with legislative requirements.
- Competitive edge over non-certified businesses.
- Improved management of environmental risk.
- Increased credibility that comes from independent assessment.
- Savings from reduced non-compliance with environmental regulations.
- Heightened employee, shareholder and supply chain satisfaction and morale.
- Meeting modern environmental ethics.
- Streamlining and reducing environmental assessments and audits.
- Increased resource productivity.

Limitations

- Prescriptive environmental performance levels are not included within the standard.
- Improvements in environmental performance can be negligible.
- Lack of public reporting, unlike other internationally recognised management systems.
- Inconsistency of ISO auditors.
- Implementing an EMS may have high cost implications for small and medium-sized enterprises.

(Only three benefits and three limitations were required.)

Suggested Answers to Study Questions

Element 3: Assessing Environmental Aspects and Impacts

Question 1

'Environmental aspect' can be defined as an "element of an organisation's activities or products or services that interacts or can interact with the environment".

'Environmental impact' can be defined as "change to the environment, whether adverse or beneficial, wholly or partially resulting from an organisation's environmental aspects".

Question 2

Life-Cycle Analysis (LCA) is a tool to identify and measure the environmental impact of a product or service throughout its life cycle, from cradle to grave. This information can then be used to inform the decision-making process regarding new products, or to make changes to materials used in existing products.

The ISO 14040 series provides guidance on aspects of LCA.

This cradle-to-grave approach helps to identify the total impact a product has on the environment.

Question 3

Direct impacts:

- Air emissions from running vehicle engines in the workshop.
- Air emissions from a waste oil burner used to provide heat for the workshop.
- Accidental spillage of oil into the public sewer at the garage.
- Contamination of the land through spillages of oil and fuel in the workshop.

Indirect impacts:

- Methane gas generated from a landfill site that disposes of the garage waste (the direct impact is caused by the company operating the landfill).
- Use of electricity from the energy provider to heat offices (the direct impact is caused by the producer of the electricity, e.g. a nuclear power plant).
- Purchasing of office desks, made from non-renewable hardwoods (the direct impacts are from emissions caused by the manufacturer of the furniture and the loss of natural resources caused by the timber company).
- Water discharges caused in the manufacture of engine parts (the direct impact is caused by the emissions from the manufacturer of the engine parts).

Suggested Answers to Study Questions

Question 4

Internal Sources	External Sources
Inspection/audit reports	Manufacturers' data
Incident data and investigation reports	Legislation
Maintenance records	Enforcement bodies' guidance documents
Job/task analysis	Government-supported organisations
Monitoring results	Trade associations and professional institutions
Raw material usage and supply	International/British Standards
Permits and consents	Commercial organisations/encyclopaedias/ textbooks/ academic journals

(Only three of each were required.)

Question 5

The following criteria might be used in deciding whether an environmental impact is significant or not:

- Scale and severity and probability/likelihood of occurrence.
- Legal or contractual requirements.
- Insufficient data or information.
- Costs of raw materials or energy.
- The concern of interested parties (views of local communities and other stakeholders) and the effect on public image.
- Sensitivity of receiving environment.
- Frequency (intermittent or continuous impact).
- Duration of impact (temporary or permanent).

(Only three were required.)

Suggested Answers to Study Questions

Element 4: Planning for and Dealing with Environmental Emergencies

Question 1

The following are some of the reasons why organisations should have emergency plans in place:

- General responsibility not to pollute the environment.
- As part of an Environmental Management System (EMS).
- To ensure prompt action to protect people and the environment.
- Risks of prosecution and other costs.
- Reputational issues.

(Only three were required.)

Question 2

The objective of testing the emergency plan should be to give confidence in the following constituents of the plan:

- The completeness, consistency and accuracy of the emergency plan and other documentation used by organisations responding to an emergency.
- The adequacy of the equipment and facilities, and their operability, especially under emergency conditions.
- The competence of staff to carry out the duties identified for them in the plan, and their use of the equipment and facilities.

Question 3

Information should include:

- The layout of the site drainage system showing clearly the difference between drains going to foul sewer and those going to surface water.
- Assembly points for staff and visitors.
- Access routes and assembly points for emergency services.
- The location of any fire hydrants or high-pressure water points.
- The location of any flammable or explosive chemicals stored on site and any other locations that may prove dangerous to emergency service personnel.

(Only three were required.)

Question 4

The three main hazards common to almost all fires are air pollution, water pollution and land pollution.

Suggested Answers to Study Questions

Element 5: Control of Emissions to Air

Question 1

The short-term effects of human exposure to air pollution include irritation and inflammation of the airways, eyes and mouth.

Question 2

Vapours are the gaseous state of materials which are liquid at normal temperature and pressure.

Mists are fine liquid droplets, usually nucleated by a particle.

Fumes refer to very small particles of less than one micron that are suspended in flue gases and air.

Question 3

Control Hierarchy	Example
Eliminate	Replace solvent-based chemicals with water-based chemicals. This has been done very effectively in the paint and ink industries where solvent-based products are now much less common than they used to be.
Minimise	This has been achieved in the motor industry through the use of improved technology, such as engine management systems and fuel injection.
Render Harmless	Fabric filters removing dust from a gas stream by passing through a fabric.

Question 4

The principle of all wet scrubbers is that water droplets are generated within the device and particles are captured within the droplets. The droplets are then removed from the air stream which is now clean. The droplets are collected as contaminated water, and transported out of the device for treatment or disposal. Wet scrubbing is used to control sticky emissions which may block filter-type collectors, to handle waste gas streams containing both particulates and gases, to recover soluble dusts and powders and to remove metallic dusts, such as aluminium, which may explode if handled dry.

Wet scrubbing techniques are normally used where the:

- Contaminant cannot be removed easily in a dry form.
- Waste gas stream contains both particulates and soluble gases.
- Particulates to be removed are soluble or wettable; they would adhere to the inner surfaces of a cyclone or bag filter plant and clog it.
- Contaminant will undergo some subsequent wet process, such as sedimentation, wet separation or neutralisation.
- Pollution control system must be compact.
- Particulates may ignite or explode if collected in a dry form.

Question 5

The main objective of LEV is to extract the flow of air away from a work process using hazardous airborne substances. The air is cleaned, often with a bag filter, before exhausting it to the outside atmosphere.

Suggested Answers to Study Questions

Element 6: Control of Environmental Noise

Question 1

The following could all be sources of noise nuisance:

- Noise from commercial activities.
- Transport noise.
- Agricultural noise.
- Construction noise.
- Quarrying and mining.
- Noise from pubs and clubs.
- Neighbour noise.
- Intruder and vehicle alarms.
- Wind farms.

(Only five were required.)

Question 2

Examples of management controls in relation to noise are:

- Control of working hours - usually to reasonable daytime hours.
- Controlling the use of radios (both music and two-way radios) - radios used for communication and entertainment can cause a nuisance to others nearby.
- Public address systems - should be designed so that sound is directed where it needs to be heard and not beyond the boundaries.
- Vehicle routes - proper routeing of vehicles, together with signage indicating any prohibited areas or routes, can reduce the nuisance caused.
- Loading doors and shutters - ensuring these are kept closed when not in use, especially during the night, can significantly reduce the noise levels and the potential for nuisance.

(Only three were required.)

Suggested Answers to Study Questions

Element 7: Control of Contamination of Water Sources

Question 1

We rely on abstracting water from many different water bodies, such as rivers and from underground, for drinking, agriculture and commercial use. Many species of wildlife rely on good quality water in rivers, streams and lakes to survive and breed. Clean water is critical to both humans and wildlife. The water cycle shows us that water is continuously moved around, in one form or another. Pollution can also be transported with the water in the same cycle. For example, a polluting discharge going into a river is likely to reach a lake. Spillage of a chemical into a lake will ultimately pollute groundwater and soil. Pollution and its harmful effects can therefore become widespread, whichever of the water bodies is polluted. All waters must therefore be protected.

Question 2

The two main categories of water pollution sources are:

- Point sources - distinct sources, such as pipelines, ditches, etc. and relatively easy to identify and control.
- Non-point or diffuse sources - including run-off from fields of fertilisers and pesticides and acid rain. They are more difficult to identify than point sources and therefore harder to control.

Question 3

The main sources of water pollution include:

- Surface water drainage - collects rainwater falling on a variety of surfaces and will wash into the system any contaminants on the surface where rain has fallen. These will then be washed into the watercourse.
- Risks of contamination from spills - many industrial sites will have a combination of foul water drains and surface water drains. It is essential that these are identified, as any spills must be contained and the appropriate regulator informed if there is a risk that the pollution will enter either a controlled water or a sewerage system.
- Process and cooling water - water is often used as a coolant and so will also collect heat. Warm water retains much lower levels of oxygen than cold water and so volumes must be controlled in order to reduce any damage to the natural environment.
- Sewage - many sewage works have storm-water systems that allow the discharge of raw sewage to a river in the event of high rainfall. Other failures in the sewerage system, such as the blocking or breaking of sewer pipes, can lead to contamination.
- Solids - grit and plastics, etc. end up in rivers, lakes and on beaches. Grits and silts (e.g. cement) are washed from building activities into rivers.
- Contamination from natural minerals, e.g. arsenic, which has been identified in groundwater in Bangladesh.

(Only three were required.)

Question 4

The main methods used to reduce contamination of water resources are:

- Screening.
- Solids separation and removal of organic load (coagulation).
- Centrifugal separation.
- Sedimentation/flotation.
- Filtration.
- Correction of pH.

(Only five were required.)

Suggested Answers to Study Questions

Element 8: Control of Waste and Land Use

Question 1

The Waste Framework Directive (2008/98/EC) describes waste from the point of view of the person discarding it as: "any substance or object which the producer or the person in possession of it discards, or intends or is required to discard".

Question 2

Under the Waste Framework Directive, a waste is classified as 'hazardous' if it meets at least one of the following criteria:

- Explosive, flammable and oxidising substances.
- Irritants and corrosives.
- Biohazards (infectious, carcinogenic, mutagenic, teratogenic).
- Ecotoxics.
- Waste that releases toxic gases in contact with water, air or an acid.
- Sensitisers.

Question 3

The five basic elements of the waste hierarchy are:

- Prevention.
- Preparing for re-use.
- Recycling.
- Other recovery.
- Disposal.

Question 4

Waste could potentially escape from control in a number of ways, including:

- Corrosion or wear of containers.
- Accidental spills or leaks.
- Breach of containment by weather.
- Blowing away or falling from vehicles or storage.
- Scavenging by vandals, thieves, children, trespassers or animals.

(Only three were required.)

Suggested Answers to Study Questions

Question 5

The following nuisances must be adequately controlled:

- **Noise nuisance**: heavy vehicle movements, both site traffic bringing waste for disposal and site plant. Licence conditions may specify working hours.
- **Odours**: minimised by the cell method of filling, ensuring the surface is covered with inert fill at the end of each working day, and operating to minimise exposed areas. Chemical sprays to mask smells may be used in unusual wind conditions.
- **Dust and litter**: minimised by damping down and good site practice, e.g. cell filling by tipping down a gradual slope and using specially designed plant to bury litter and increase the fill density, maximising the site capacity and minimising later settlement. This also reduces the possibility of fire (producing smoke, smells and contaminated leachate) starting in the waste.
- **Vermin**: gulls, rats, mice and foxes. Good site practice is required, and possibly an eradication programme for rats.

Question 6

	Advantages	Disadvantages
Landfill	<ul><li>Comparatively cheap disposal.</li><li>Can be used to restore areas used for quarries, etc. to local amenity use.</li><li>Able to take large volumes of waste.</li><li>Modern landfill is of a low risk for humans and the environment.</li><li>Can be used to generate electricity.</li></ul>	<ul><li>Cheap, so does not encourage producers and consumers to migrate up the waste hierarchy.</li><li>Waste does not break down quickly, so the problem of management remains for a considerable time after a site has closed.</li><li>Poorly managed sites can lead to:<ul><li>Surface and groundwater pollution from leachate.</li><li>Air pollution from unmanaged LFG or fires.</li></ul></li></ul>
Incineration	<ul><li>Reduction in volume.</li><li>Allows heat to be recovered.</li><li>An energy source.</li></ul>	<ul><li>High initial cost.</li><li>Volume of traffic.</li><li>Monitoring of air pollution.</li><li>Disincentive to recycling schemes (incinerators may need constant feeding).</li></ul>

(Only three advantages and three disadvantages of each were required.)

Question 7

Direct water contamination could occur by migration through plastic water pipes (especially of phenols and cresols) into household water supplies. Indirect water contamination could occur by leaching into groundwater.

Suggested Answers to Study Questions

Question 8

The hazards associated with land contamination include:

- Ingestion.
- Food chain plant uptake.
- Contamination of drinking water.
- Prevention/inhibition of plant growth.
- Odours.
- Fumes.
- Fire and explosion.
- Direct contact.
- Building damage.

Suggested Answers to Study Questions

Element 9: Sources and Use of Energy and Energy Efficiency

Question 1

Some of the advantages and disadvantages of the use of fossil fuels are:

Advantages	Disadvantages
Straightforward combustion process.	Major contributor to climate change.
Relatively inexpensive.	Cause acid rain.
Easily transported.	Use a non-renewable source of energy so not sustainable in the long term.
Large amounts of electricity can be generated in one place, quite cheaply.	Prices are susceptible to changes in global politics so may rise significantly at short notice.
Gas-fired power stations relatively efficient.	Extracting raw materials can be dangerous and damaging to the environment.
Power stations can be built almost anywhere.	Emissions may contribute to poor air quality locally, thereby affecting people's health.

(Only three advantages and three disadvantages were required.)

Question 2

The three main ways in which we use solar power are:

- **Solar cells** (photovoltaic or photoelectric), where photovoltaic panels convert light into electricity at an atomic level through the use of materials that exhibit a property known as a photoelectric effect. This effect causes them to absorb photons of light and release electrons. When these electrons are captured, an electric current is generated.
- **Solar water heating**, where energy from the Sun is used to directly heat water in glass panels, thereby reducing the amount of energy from fossil fuels required to provide hot water for use in the house.
- **Solar furnaces**, which are commercial installations that use a large number of mirrors to concentrate the energy of the Sun into a small space and allow the production of very high temperatures. Some of these furnaces can produce temperatures up to 33,000°C.

Question 3

Combined Heat and Power (CHP) is the generation of usable heat and power (usually electricity) in a single process.

CHP systems can be employed over a wide range of sizes, applications, fuels and technologies. In its simplest form, CHP employs a gas turbine, engine or steam turbine to drive an alternator and the resulting electricity can be used either wholly or partially on site with any excess being supplied to the national grid system. The heat produced during power generation is recovered, usually in a heat recovery boiler, and can be used to raise steam for a number of industrial processes, to provide hot water for space heating or, with appropriate equipment installed, cooling.

CHP is a form of decentralised energy technology. CHP systems are typically installed on site, supplying customers with heat and power directly at the point of use and therefore helping to avoid the significant losses which occur in transmitting electricity from a large centralised plant to the customer.

Question 4

'Peak load' and 'peak demand' are two terms that are used interchangeably to denote the maximum power requirement of a system at a given time, or the amount of power required to supply customers at times when need is greatest.